The Play of Animals

THE PLAY OF ANIMALS

BY

KARL GROOS

PROFESSOR OF PHILOSOPHY IN THE
UNIVERSITY OF BASEL

TRANSLATED WITH THE AUTHOR'S CO-OPERATION

BY ELIZABETH L. BALDWIN

WITH A PREFACE AND AN APPENDIX BY

J. MARK BALDWIN

PROFESSOR IN PRINCETON UNIVERSITY

1898

NEW YORK

D. APPLETON AND COMPANY

EDITOR'S PREFACE.*

In this volume Professor Groos makes a contribution to three distinct but cognate departments of inquiry: philosophical biology, animal psychology, and the genetic study of art. Those who have followed the beginnings of inquiry into the nature and functions of play in the animal world and in children will see at once how much light is to be expected from a thorough-going examination of all the facts and observations recorded in the literature of animal life. This sort of examination Professor Groos makes with great care and thoroughness, and the result is a book which, in my opinion, is destined to have wide influence in all these departments of inquiry.

I wish, before speaking of certain conclusions which are of especial interest, to make some running comments on the contents of the book, without, of course, forestalling the reader's own discovery of its riches. Chapter I is an examination of Mr. Spencer's "surplus-energy" theory of play; the result of which is, it seems, to put this theory permanently out of court. The author's

* In this preface certain passages are repeated from a review of the German edition of Professor Groos' book, printed in Science, February 26, 1897.

main contention is that play, so far from being "by-play," if I may so speak, is a matter of serious moment to the creature. Play is a veritable instinct. This view is expanded in Chapter II, where we find a fine treatment in detail of such interesting topics as imitation in its relation to play, the inheritance of acquired characters apropos of the rise of instincts, and the place and function of intelligence in the origin of these primary animal activities. This chapter, dealing with the biological theory of play, is correlated with Chapter V, in which the Psychology of Animal Play is treated. Together they furnish the philosophical and theoretical basis of the book, as the chapters in between furnish the detailed data of fact. I shall return to the biological matter below. Chapters III and IV go into the actual Plays of Animals with a wealth of detail, richness of literary information, and soundness of critical interpretation which are most heartily to be commended. Indeed, the fact that the first book on this subject is at the same time one of such unusual value, both as science and as theory, should be a matter of congratulation to workers in biology and in psychology. The collected cases, the classification of animal plays, as well as the setting of interpretation in which Professor Groos has placed them—all are likely to stand, I think, as a piece of work of excellent quality in a new but most important field of inquiry.

With this general and inadequate notice of the divisions and scope of the book, I may throw together in a few sentences the main theoretical positions to which the author's study brings him. He holds play to be an instinct developed by natural selection (he gives good reasons for not accepting the inheritance of acquired characters), and to be on a level with the other instincts which are developed for their utility. It is very near, in its origin and function, to the instinct of imitation, but yet they are distinct (a word more below on the relation between play and imitation). Its utility is, in the main, twofold: First, it enables the young animal to exercise himself beforehand in the strenuous and necessary functions of its life and so to be ready for their onset; and, second, it enables the animal by a general instinct to do many things in a playful way, and so to learn for itself much that would otherwise have to be inherited in the form of special instincts; this puts a premium on intelligence, which thus comes to replace instinct (p 71). Either of these utilities, Professor Groos thinks, would insure and justify the play instinct; so important are they that he suggests that the real meaning of infancy is that there may be time for play (see his preface). This general conception of play has been set forth by other writers; but Professor Groos works it out in this book in a way which attaches his name permanently to it.

It is especially in connection with this latter function of play, I may add, that the instinct to imitate comes in to aid it. Imitation is a real instinct, but it is not always playful; play is a real instinct, but it is not always imitative. There is likely, however, to be a great deal of imitation in play, since the occasion on which a particular play-function develops is often that which also develops the imitative tendency as well—i. e., the actual sight or hearing of the acts or voices of other animals. Moreover, the acquisition of a muscular or vocal action through imitation makes it possible to repeat the same action afterwards in play.

It is only a step, therefore, to find that imitation, as an instinct, has to have ascribed to it, in a measure, the same race utility as play—that of going before the intelligence and preparing the way for it, by rendering a great number of specialized instincts unnecessary. It is interesting to the present writer to contrast this view with that which he has himself recently developed *— i. e., the view that imitation supplements inadequate congenital variations in the direction of an instinct, and so, by keeping the creature alive, sets the trend of further variations in the same direction until the instinct is fully organized and congenital. If both of these views be true, as there seems reason to believe, then imitation holds a remarkable position in relation to intelligence and instinct. It stands midway between them and aids them both. In some functions it keeps the performance going, and so allows of its perfection as an instinct; in others it puts a stress on intelligence, and so allows the instinct to fall away, if it have no independent utility in addition to that served by the intelligence.† In other words, it is through imitation that instincts both arise and decay; that is, some instincts are furthered, and some suppressed, by imitation. And all this is accomplished with no appeal to the inheritance of acquired characters, Professor Groos agreeing with Weismann that the operation of natural selection as generally recognised is probably sufficient (see his preface). For myself I find most helpful the theory of Organic Selection referred to by Professor Groos on pages 64 and

* See Science, March 20, 1896.

† In a private communication Professor Groos suggests to me that the two views may well be held to supplement each other. The case is very much the same with early intelligence, in the form of Association of Ideas: where it fully accomplishes the utility also subserved by an instinct, it tends to supersede the instinct; otherwise it tends to the development of the instinct (Groos, this edition. p. 71, and Baldwin, Science, April 10, 1896).

65. Following up his kindly reference, I venture, with his concurrence, to reprint as an Appendix to this translation a short article of my own on Organic Selection.

The difficulty which I see to this conception of play as a pure instinct is that which is sometimes urged also against considering imitation an instinct—i. e., that it has no definite motor co-ordinations, but has all the variety which the different play-forms show. If the definite congenital plays are considered each for itself, then we have a great many instincts, instead of a general play instinct. But that will not do, for it is one of Professor Groos' main contentions, in the chapter on The Psychology of Animal Play, that they have a common general character which distinguishes them from other specialized instinctive actions. They are distinguished as play actions, not simply as actions. This difficulty really touches the kernel of the matter, and serves to raise the question of the relation of imitation to play; for imitation presents exactly the same conditions—a general tendency to imitate, which is not exhausted in the particular actions which are performed by the imitation. I shall remark on the solution of it below, in speaking of Professor Groos' psychology of play. It will be interesting to see how he treats this problem in his promised work on Die Spiele der Menschen, for the imitative element is very marked in children's plays In view of this objection to the use of the term "instinct" for play—"impulse" possibly being better—I venture to suggest that the theory which regards play as a native tendency of the animal to practise certain special functions, before they are really required of him, be called the "practice theory" of play.

Other matters of interest in this biological part are the great emphasis which Groos finds it necessary to put on "tradition," instruction, imitation, etc., in young animals, even in enabling them to come into possession of their natural instincts; in this the book tends in the same direction as the later volume of Prof. C. Lloyd Morgan. The present writer has also emphasized the fact under the term social heredity. Again, there is an acute discussion of Darwin's Sexual Selection, a discussion which Professor Groos sufficiently explains in his own preface.* I find an anticipation of the position—as it were, a happy intuition—in the Non-Religion of the Future of M. Guyau (page 302. Again, the imperfect character of most instincts is emphasized, and the interaction with imitation and intelligence.

Finally, I should like to suggest that a possible category of "Social Plays" might be added to Groos' classification; plays in which the utility of the play instinct seems to have reference to social life as such. In such a category it might be possible to place certain of the animals' performances which seem a little strained under the other heads; and also those performances in which the social function of *communication* is playfully exercised. A good deal might be said also in question of the author's treatment of "Curiosity" (*Neugier*). He makes curiosity a matter of the attention, and finds the restless activity of the attention a play function. My criticism would be that while curiosity does bring the animal into possession of the details of knowledge before they are pressed in upon him by harsh experience, yet attention does not altogether fulfil the requirements of the author's psychological theory of play.

---

* "Sexual" is referred back to "natural" selection. although the direct results of such preferential mating would still seem to be a "determination" of variations for natural selection to work upon (*cf.* Science, November 23, 1896, p. 726).

Turning now to the interesting question of the psychological theory, we find it developed, as it would have to be, in a much more theoretical way. The play consciousness is fundamentally a form of "conscious self-illusion"—*bewusste Selbsttauschung* It is just the difference between play activity and strenuous activity that the animal knows, in the former case, that the situation is not real, and still allows it to pass, submitting to a pleasant sense of "make-believe." It is only fair to say, however, that Herr Groos admits that in certain more definitely instinctive forms of play this criterion does not hold; it would be difficult to assume any consciousness of self-illusion in the fixed courting and pairing plays of birds, for example. The same is seen in the very intense reality which a child's game takes on sometimes for an hour at a time. Indeed, the author distinguishes four stages in the transition from instincts in which the conscious illusion is absent, to the forms of play to which we can apply the phrase "play activity" in its true sense—i. e., that of *Scheinthatigkeit*. The only way to reconcile these positions that I see is to hold that there are two different kinds of play: that which is not psychological at all—i. e., does not show the psychological criterion at all—and that which is psychological as "conscious self-illusion." Herr Groos does distinguish between "objective" and "subjective" *Scheinthatigkeit* (p. 292). The biological criterion of definite instinctive character might be invoked in the former class, and the psychological criterion in the other; and we would then have a situation which is exemplified in many other functions of animal and human life—functions which are both biological and instinctive, and also psychological and intelligent, as, e. g., sympathy, fear, bashfulness. Then, of course, the further question comes up as to which of these forms is primary; again the old problem as to whether intelligence arose out of reflexes or the reverse.

I think some light falls on this time-honoured problem from the statement of it in connection with this new question of play; especially when we remember Herr Groos' theory of the function of imitation with the extension of his view suggested above. If imitation stands midway between instinct and intelligence, both furthering the growth of instinct in some cases, and also, in other cases, leading to its decay in the presence of intelligence, then we might hold something like this: In proportion as an action loses its consciously imitative and volitional character, to that degree it tends to be incapable of "make-believe" exercise, becoming real in consciousness and instinctive in performance (and this applies to the cases in which imitation has itself become habitual and instinctive, as in the mocking-bird); and on the contrary, in proportion as an instinctive action is modified and adapted through imitation and intelligence, to that degree it becomes capable of assuming the "make-believe" character and is indulged in as conscious play. I can not enlarge upon this here, but it seems to square with a good many of the facts; both with those which Professor Groos cites as showing that imitation opens the way for the decay of instinct with the growth of intelligence, and with those which Professor Lloyd Morgan and I have cited as showing that imitation keeps congenital variations alive and so allows them to accumulate into instincts. It is also consistent with the view that imitation is a sort of meeting point of race habit, represented by instinct, and race accommodation, represented by intelligence: just the double function which imitation serves also in the development of the individual, as I have argued in detail in my volume on Mental Development in the Child and the Race.

Going into the analysis of the play psychosis, Herr Groos finds several sources of pleasure to the animal in it: pleasure of satisfying an instinct, pleasure of movement and energetic action, but, most of all, "pleasure in being a cause." This last, together with the "pleasure in experimenting," which characterizes many play activities, is urged with great insistence, and properly so. Even the imitative function is said to produce the joy of "victory over obstacles." Yet here again the author is compelled to draw the distinction between the play which is psychological enough to have a represented object, and the instinctive sort in which the pleasure is only that of the instinct's own performance. The pleasure of overcoming friction of movement, also, seems very doubtful, since in most games we stop playing when the friction and inertia of the muscles come to consciousness as fatigue. Much more, however, is to be said for the pleasure of rivalry, or of overcoming an opponent, in the higher types of play; but Herr Groos scarcely does this justice.

The second element in the play or *Schein* consciousness is the feeling of freedom (*Freiheitsgefühl*). In play there is a sense of "don't-have-to," so to speak, which is contrasted both with the necessity of sense and with the imperative of thought and conscience. This idea seems to be part of Schiller's theory of play. So Groos thinks the general feeling of freedom holds in consciousness only while there is a play of motives, to which the agent may put an end at any moment—a sense of "don't-have-to" in the life of choice. This sense of freedom keeps the "make-believe" consciousness pure and prevents our confusing the game with the real activities of life This is very interesting and suggestive. The sense of freedom is certainly prominent in play. Whether it should be identified with the sense of control which has been used by some writers as a criterion (both in a negative and in a positive sense) of the belief in realities already experienced, or again with the freedom with which choice is pregnant, is more questionable. Without caring to make a criticism of Professor Groos' position, I may yet point out that in our choices there are those which are free with a "don't-have-to" freedom, and there are choices—and these are the momentous ones, the ones to which freedom that men value attaches—which are strenuous and real in the extreme. Indeed, it seems paradoxical to liken the moral life, with its sense of freedom, to a "game of play," and to allow the hard-pressed sailor on the ethical sea to rest on his oars behind a screen of *Schein* and plead, "I sha'n't play." Seriously, this is what some other writer might press on to; and it comes out again in the author's extremely interesting sections on art, of which I may say a word in conclusion.

Those who have read Professor Groos' former stimulating book, Einleitung in die Æsthetik, will anticipate the connection which he finds between play and art. The art consciousness is a consciousness of an "inner imitation," which is in so far "make-believe" as contrasted with reality. The "self-conscious illusion" of the play consciousness is felt in extreme form in the theatre, and it is found to be pleasurable even when we play with painful situations, as in tragedy In art the desire to make an impression on others shows the "pleasure of being cause." This intent to work on others is a necessary ingredient in the art impulse. Groos differs from K. Lange, who holds a similar view of the necessary division of consciousness between reality and "make-believe" in the æsthetic, in that Lange thinks there must be a continual oscillation between the two poles of the divided consciousness, while Groos thinks there is rather a settling down in the state of illusion, as in an artist's preoccupation with his creations, a novelist with his characters, and a child with her doll. In art the other great motive of play, "experimenting," is also prominent, and is even more fundamental from a genetic point of view.

Here again the question left in my mind is this: whether the "make-believe" motive is really the same as the art motive. Do we not distinguish between the drama (to take the case most favourable to the theory) as amusement and the drama as art? And does the dramatist who is really an artist write to bring on a *consciousness* of self-illusion in the spectator by presenting to him a "make-believe" scene? Does he not rather aim to produce an "inner imitation" in him which shall arouse the emotional and volitional attitudes of full reality? There does seem to be, in a work of fine art, a strenuous outreach not only toward the imitation of truth, but toward the actual *conviction* of truth. It may be that we should distinguish with Aristotle between truth which comes to us didactically and truth which comes artistically, and find in the method of the latter, and in that alone, the source of æsthetic impression; but even then we should not have to feel the æsthetic creation to be "make-believe." In any case the theory of Professor Groos, which has its roots in the views of Lange and Von Hartmann, is extremely interesting and valuable, especially as contrasted with the recent psychological theory of Mr. H. R. Marshall. As to Professor Groos' theory, musical art would present difficulties, and so would lower sensuous æsthetic effects generally.

Genetically art rests upon play, according to Herr Groos, in that the three great motives of art production, "Self-exhibition" (*Selbstdarstellung*), "Imitation," and "Decoration" (*Ausschmuckung*), are found in the three great classes of animal plays, respectively, "Courting," "Imitation," and "Building Art" (*Baukünste*, seen in birds' nest-building, etc.). On the strength of this, Professor Groos finds both æsthetic appreciation and impulse in the animals, and all rests upon the original "experimenting" impulse of play. Of this, however, Professor Groos does not give a satisfactory account, I think. Experimenting seems to be a necessary part of effective learning by "imitation," and the use made of it in the selection of movements may be its original use. I have suggested elsewhere (Social and Ethical Interpretations, sections 98 to 102) some reasons for thinking also that decorative art may have sprung from the "self-exhibiting" impulse, thus reducing the æsthetic motives to two.

On the whole, Professor Groos' book is both a pioneer work and one of great permanent value. In venturing to criticise it I have thought it best to raise points of discussion—even though to a thinker like Professor Groos they may be trivial and easily answered—as fitted to give to the lay reader a sense of the larger issues for the sake of which, after all, the delightful stories of animal life in the book have been collected by the author.

J. MARK BALDWIN.

PRINCETON, *April, 1898.*

The translator wishes it to be said that all the alterations made for this edition have been either requested or approved by the author, only some few footnotes of a bibliographical sort having been added after Professor Groos saw the proofs. The additions made by the translator in the notes are put in brackets, both those which Professor Groos has seen and also the very few others which were subsequently added. The reader will also notice from the title-page that the author has now been called from his former position at Giessen to the chair in Philosophy in the University of Basel.

J. M. B.

AUTHOR'S PREFACE.

ANIMAL psychology is regarded by many somewhat contemptuously as a sort of amusement, from which nothing worth speaking of can be expected for the advancement of our modern science of the mind. I do not believe this. In the first place, it is quite wrong to judge animal psychology mainly from its value for the interpretation of the mind of man, making secondary the independent interest to which it lays claim. Yet, apart from this, such a study is valuable to the anthropologist in many ways, though it must be admitted that but little has as yet been accomplished in this direction. Unfortunately, many of the works hitherto published on the subject of animal psychology labour under the disadvantage of being strongly biased, and suffer also from a lack of method. Their authors, justly indignant at the arrogance of man in despising the animals and claiming for themselves all the higher and more refined attributes, naturally wish to prove that animals, too, possess a high degree of intelligence and feeling; they accordingly emphasize the resemblance of animals to man, and their work becomes an interesting collection of anecdotes of specially gifted individual animals—collections, no doubt, possessing much intrinsic worth but of little value to the psychologist. If the observation of animals is to be rendered fruitful for the unsolved problems of anthropology, an untried way must be entered upon; attention must be directed less to particular resemblances to man, and more to specific animal characteristics. Hereby a means may be found for the better understanding of the animal part in man than can be attained through the discussion of human examples alone. Man's animal nature reveals itself in instinctive acts, and the latest investigators tell us that man has at least as many instincts as the brutes have, though most of them have become unrecognisable through the influence of education and tradition. Therefore an accurate knowledge of the animal world, where pure instinct is displayed, is indispensable in weighing the importance of inherited impulses in men.

The number of investigators who have adopted this method is not great, and I venture to hope that this book may be in some degree influential in increasing it, as well as respect for animal psychology as a science.

The world of play, to which art belongs, stands in most important and interesting contrast with the stern realities of life; yet there are few scientific works in the field of human play, and none at all in that of animal play—a fact to be accounted for, probably, by the inherent difficulties of the subject, both objective and subjective. The animal psychologist must harbour in his breast not only two souls, but more; he must unite with a thorough training in physiology, psychology, and biology the experience of a traveller, the practical knowledge of the director of a zoölogical garden, and the outdoor love of a forester. And even then he could not round up his labours satisfactorily unless he were familiar with the trend of modern æsthetics. Indeed, I consider this last point so important that I venture to affirm that none but a student of æsthetics is capable of writing the psychology of animals. If in this statement I seem to put myself forward as a student of æsthetics, I can only say that I hope for indulgence, in view of the many shortcomings which are apparent in this effort, on the ground that a versatility so comprehensive is unattainable by an ordinary mortal.

The first two chapters seek to establish the conception of play on a basis of natural science. There are two quite different popular ideas of play. The first is that the animal (or man) begins to play when he feels particularly cheerful, healthy, and strong; the second—which I found even entertained by a forester—that the play of young animals serves to fit them for the tasks of later life. The former view tends to a physiological, the latter to a biological, conception of play. The first finds its scientific basis in the theory of surplus energy, which is amplified by Herbert Spencer especially, but which was previously promulgated by Schiller, as I have attempted to show in the beginning of the book. This explanation of play is certainly of great value, but is not fully adequate, and I have reached the conclusion that a state of surplus energy may not always be even a *conditio sine qua non* of play.

The physiological conditions which cause a young beast of prey to follow a rolling ball need not, apparently, be different from those of the grown animal in pursuit of its natural prey. The other view, by keeping before the eyes the biological significance of play, seems to me to open the way to a more thorough understanding of the problem.

This reference to biology brings me at once to the difficult question of instinct. After a long historical and critical investigation of the subject, I intrench myself in the principles defined by H. E Ziegler, who, as a disciple of Weismann, refers all instincts directly to natural selection. Accordingly, I have not used the Lamarckian principle of the transmission of acquired characters, which, to say the least, is doubtful, in the interpretation of fact. On this definition of instinct as a basis a new biological theory of play is developed, of which the following are the main points: The real problem lies in the play of the young; that once successfully explained, adult play will offer no special difficulties. The play of youth depends on the fact that certain instincts, especially useful in preserving the species, appear before the animal seriously needs them. They are, in contrast with later serious *exercise* (*Ausubung*), a *preparation* (*Vorübung*) and *practice* (*Einübung*) for the special instincts. This anticipatory appearance is of the utmost importance, and refers us at once to the operation of natural selection; for, when the inherited instinct may be supplemented by individual experience, it need not be so carefully elaborated by selection, *which accordingly favours the evolution of individual intelligence as a substitute for blind instinct.* At the moment when the intelligence reaches a point of development where it is more useful than the most perfect instinct, natural selection will favour individuals in whom instinct appears only in an imperfect form, manifesting itself in early youth in activity purely for exercise and practice—that is to say, *in animals which play.* Indeed, the conclusion seems admissible, in summing up the biological significance of play, that perhaps *the very existence of youth is due in part to the necessity for play;* the animal does not play because he is young, he has a period of youth because he must play. Whoever has observed the tremendous force of the play impulse in young animals will hardly fail to give this thought some hospitality.

Though calling in the principle of natural selection exclusively, on the lines of Weismann's theory, in explaining these phenomena, I am by no means convinced of the all-sufficiency of this law, but freely admit the possibility that still other and perhaps unknown forces contribute their influence in this process of evolution. The conception of evolution itself is gaining strength and assurance with the progress of time, but with respect to specific Darwinism a note of *fin-de-siècle* lightness is audible to the attentive ear. I do not know whether the following idea has occurred to any one else, but to me it is somewhat baffling. It is quite conceivable that a man might arise and say: "Three of the most distinguished investigators in the subject of descent are Wallace, Weismann, and Galton. Now, I agree with Wallace in discarding sexual selection; I hold with Weismann that the inheritance of acquired characters is impossible; and I combat with Galton the idea that natural selection is sufficient to explain the change from an established species to a new one." What, then, is left of the Darwinian theory of organic evolution?

In the third and fourth chapters a system of animal play is developed for the first time on the biological theory as a basis. The variety and scope of such play has been up to this time very much underrated, as, I believe, this classification and grouping under important heads will show. The discussion of curiosity developed a theory of attention which was simultaneously published as a short article, Ueber unbewusste Zeitschatzung, in Zeitschrift für Psychologie u. Physiologie der Sinnesorgane. In the introduction to the chapter devoted to love plays I have attempted an essential modification of the doctrine of sexual selection, making it a special case of natural selection. While agreeing with Wallace that the unusual colours and forms, as well as complicated calls, are to be considered as largely a means of defence and offence and of recognition among animals, I yet believe that in their higher manifestations such phenomena often have a very close connection with sexual life. This is more obviously the case with the display of ornamentation, of skill in flying, dancing, and swimming, and in bird-song. The disciple of Weismann who can not accept Spencer's explanation of such phenomena must either cleave to Darwin's sexual selection, as Weismann himself does, or seek a new principle. Such a principle I believe I have found. It depends on two closely related facts. As sexual impulse must have tremendous power, it is for the interest of the preservation of the species that its discharge should be rendered difficult. This result is partly acomplished in the animal world by the necessity for great and often long-continued excitement as a prelude to the act of pairing. This thought at once throws light on the peculiar hereditary arts of courtship, especially on the indulgence in flying, dancing, or singing by a whole flock at once. But the hindrance to the sexual function that is most efficacious, though hitherto unappreciated, is the instinctive coyness of the female. This it is that necessitates all the arts of courtship, and the probability is that seldom or never does the female exert any choice. She is not awarder of the prize, but rather a hunted creature. So, just as the beast of prey has special instincts for finding his prey, the ardent male must have special instincts for subduing feminine reluctance; and just as in the beast of prey the instinct of ravenous pursuit is refined into the various arts of the chase, so from such crude efforts at wooing, that courtship has finally developed, in which sexual passion is psychologically sublimated into love. According to this theory, there is choice only in the sense that the hare finally succumbs to the best hound, which is as much as to say that the phenomena of courtship are referred at once to natural selection. It follows, too, that however useful attractive form and colouring may be in relation to other ends, they certainly contribute to that of subduing feminine coyness, and hence further the sexual life.

The last chapter treats of the psychological aspects of play. Setting out from the physiological side, I lead up to the central idea of the whole conception, namely, "*joy in being a cause*"; which seems to me to be the psychic accompaniment of the most elementary of all plays, namely, experimentation. From here as a starting point it permeates every kind of play, and has even in artistic production and æsthetic enjoyment a significance not sufficiently appreciated.

But the principal content of the closing chapter is the investigation of the more subtle psychic phenomenon that is connected with the subject, namely, "make-believe," or "conscious self-illusion." The remarks on divided consciousness and the feeling of freedom during make-believe activity prove that the attempt to penetrate into the modern æsthetic problem is a serious undertaking. They point to a field beyond the limits of the subject of this treatise, which I hope to discuss exhaustively in my next work, having human play for its subject.

KARL GROOS.

GIESSEN, *October, 1895.*

CONTENTS.

THE PLAY OF ANIMALS.

CHAPTER I.

THE SURPLUS ENERGY THEORY OF PLAY.

THE most influential theory of play explains it by means of the surplus energy principle. In what follows I shall attempt to demonstrate that this theory has not the scope usually attributed to it. It owes its development and extension principally to Herbert Spencer, but it is based on a principle of Schiller's, in whose philosophy, however, it holds but a subordinate place. It is necessary here in the beginning of the inquiry, to set Schiller's priority in the right light, as it does not seem to be generally known. Schiller's treatment of play and the play instinct is to be found in his excellent letters On the Æsthetic Education of Mankind. Later I shall enter more fully into their contents, confining myself here to the passage on which the theory of surplus energy is especially based. It is in the twenty-seventh letter, and reads as follows: "Nature has indeed granted, even to the creature devoid of reason, more than the mere necessities of existence, and into the darkness of animal life has allowed a gleam of freedom to penetrate here and there. When hunger no longer torments the lion, and no beast of prey appears for him to fight, then his unemployed powers find another outlet. He fills the wilderness with his wild roars, and his exuberant strength spends itself in aimless activity. In the mere joy of existence, insects swarm in the sunshine, and it is certainly not always the cry of want that we hear in the melodious rhythm of bird-songs. There is evidently freedom in these manifestations, but not freedom from all necessity, only from a definite external necessity. The animal works when some want is the motive for his activity, and plays when a superabundance of energy forms this motive—when overflowing life itself urges him to action." * I will not assert that in his choice of examples from animal life Schiller has here set forth particularly clear or unchallengeable cases, but that what he had to say about them is expressed with perfect clearness—namely, that the animal is impelled to serious work by an external want, but to play by his own superfluity of energy. Through the one he restores his depleted powers; by means of the other he gives vent to superfluous ones.

Jean Paul and J. E. Beneke express themselves much as Schiller does with reference to human play. "Play," says Jean Paul in Levana (§ 49), "is at first the expression of both mental and physical exuberance. Later, when school discipline has subjected all the passions to rule, the limbs alone give expression to the overflowing life by running, leaping, and exercising generally." And Beneke says, "The child directs his superfluous energy chiefly to play," † and traces this tendency back to "conservation of original powers." ‡

* See also Schiller's poem, Der spielende Knabe, first published in 1800 in the first volume of poems: "Yet exuberant strength makes its own fancied bounds."

† Erziehungs- und Unterrichtslehre, Berlin, 1835, i, 131.

‡ Lehrbuch der Psychologie als Naturwissenschaft, Berlin, 1833, p. 24.

Spencer gives a short account of his theory in the last chapter of the Principles of Psychology, which treats of the æsthetic feelings. "Many years ago," says he (§ 533), "I met with a quotation from a German author to the effect that the æsthetic sentiments originate from the play impulse. I do not remember the name of the author, and if any reasons were given for this statement, or any inferences drawn from it, I can not recall them. But the statement itself has remained with me, as being one which, if not literally true, is yet the adumbration of a truth." It is now well known to many readers of Spencer from what German work was derived this citation which made such a lasting impression on him. Many have publicly expressed themselves on the subject, as Sully, Grant Allen,* and myself in my Einleitung in die Æsthetik.† The doctrine of the origination of the æsthetic feelings from play impulses is the cardinal point of Schiller's theory of the beautiful as it is revealed to us in these letters on æsthetic education. Schiller himself, not to speak of Kant, may have been influenced by Home, and so the idea merely found its way back to England when he in turn influenced Spencer. So far this indebtedness of Spencer to Schiller is pretty generally recognised in professional circles. But it is quite otherwise with the passage just quoted; it occurs in a part of the Æsthetics letters, comparatively unfamiliar, and therefore seemingly overlooked by most readers. "The theory" (of play impulse), says Wallaschek, "remained unheeded. though committed to writing nearly a century ago. Put, in our times, into

* See R Wallaschek, On the Origin of Music, Mind, vol xvi (1891), p 376.

† P. 176.

scientific form by Mr. Herbert Spencer, it has nothing in common with its earlier presentment beyond the name, the grounds being quite different." Had the above-cited passage from Schiller's letters been known to Wallaschek, he could never have written this statement, for it sets forth in plain words the very "grounds" on which Mr. Spencer founded his theory—namely, the doctrine of superfluous energy as the cause of play. Moreover, Schiller is the forerunner of Spencer, not only in that he derives the æsthetic feelings from play impulses, but also in teaching that play impulse itself has its origin in superfluous energy. How far-reaching this correspondence is will be seen if I now let Spencer speak: "Inferior kinds of animals have in common the trait that all their forces are expended in fulfilling functions essential to the maintenance of life. They are unceasingly occupied in searching for food, in escaping from enemies, in forming places of shelter, and in making preparation for progeny. But as we ascend to animals of high types, having faculties more efficient and more numerous, we begin to find that time and strength are not wholly absorbed in providing for immediate needs. Better nutrition, gained by superiority, occasionally yields a surplus of vigour. The appetites being satisfied, there is no craving which directs the overflowing energies to the pursuit of more prey or to the satisfaction of some pressing want. The greater variety of faculty commonly joined with this greater efficiency of faculty has a kindred result. When there have been developed many powers adjusted to many requirements, they can not all act at once; now the circumstances call these into exercise and now those, and some of them occasionally remain unexercised for considerable periods. Thus it happens that in the more evolved creatures there often occurs an energy somewhat in excess of immediate needs, and there comes also such rest, now of this faculty and now of that, as permits the bringing of it up to a state of high efficiency by the repair which follows waste." If we add to this the fact that such overflow of energy is explained by Spencer physiologically as a reintegration which more than balances the using up of brain cells, thus producing in the cells an "excessive readiness to decompose and discharge," we have become acquainted with the foundation of Spencer's theory of play. It is perfectly evident that it has more in common with Schiller's theory than the mere name; that, indeed, in its "grounds" it fully coincides with the passage cited from the Æsthetics letters.* In one point only does Spencer go beyond Schiller's conception: he connects the idea of imitation with that of the overflow of energy. And it is exactly at this point that Spencer seems to me to have erred. I will return to his own text and endeavour to show that he can not substantiate his data. After he has given the foregoing physiological explanation of surplus energy, he goes on: "Every one of these mental powers then being subject to this law, that its organ, when dormant for an interval longer than ordinary, becomes unusually ready to act, unusually ready to have its correlative feelings aroused, giving an unusual readiness to enter upon all the correlative activities; it happens that a simulation of these activities is easily fallen into, where circumstances offer, in place of the real activities. Hence play of all kinds." "It is," says R. Wallaschek, in agreement

* It is of course possible that Spencer may, notwithstanding this coincidence, have arrived at the idea of surplus energy independently of Schiller.

with Spencer, "the surplus vigour in more highly developed organisms, exceeding what is required for immediate needs, in which play of all kinds takes its rise, manifesting itself by way of imitation or repetition of all those efforts and exertions which are essential to the maintenance of life." *

In review I may here enumerate the essential points so far made:

1. The higher animals being able to provide themselves with better nourishment than the lower, their time and strength are no longer exclusively occupied in their own maintenance, hence they acquire a superabundance of vigour.

2. The overflow of energy will be favoured in those cases where the higher animals have need for more diversified activities, for while they are occupied with one, the other special powers can find rest and reintegration.

3. When, in this manner, the overflow of energy has reached a certain pitch, it tends to discharge.

4. If there is no occasion at the moment for the correlative activity to be seriously exercised, simply imitative activity is substituted, and this is play.

There can be no doubt that the conception of play thus set forth is very plausible, but its inadequacy can easily be demonstrated. Should play indeed be universally considered as the imitation of serious activities, for which there may be inclination but no opportunity? There is, of course, no doubt that imitation is of the greatest importance in much play, and I shall often have occasion to refer in the sequel to the imitative impulses. Nevertheless it is true that the conception of

* On the Origin of Music, Mind, xvi (1891), p. 376.

imitation here set forth—namely, as the repetition of serious activities to which the individual has himself become accustomed—can not be applied directly to the primary phenomena of play—that is, to its first elementary manifestations, to the play of young animals and of children. For such plays, which must be explained at the very outset in order to get a satisfactory conception of the subject, are very often not *imitations* (*Nachahmungen*), but rather *premonitions* (*Vorahmungen*) of the serious occupations of the individual. The "experimenting" of little children and young animals, their movement, hunting, and fighting games, which are the most important elementary forms of play, are not imitative repetitions, but rather preparatory efforts. They come before any serious activity, and evidently aim at preparing the young creature for it and making him familiar with it. The tiny bird that tries its wings while still in the nest; the antelope that (as Dr. A. Seitz, director of the zoölogical gardens at Frankfort, tells me) attempts to practise leaping at the age of six weeks; the young monkey that playfully seizes anything within his reach, and is only quieted when he has caught his claws in the tufts of hair on his own body, and fettered them; the giraffe that is at home in its cage by the third day of life; the feline tribe that learn so early to cling by their claws; the dog which educates itself, by play, for fighting with other dogs, and for pursuing, seizing, shaking, and rending its prey; the infant that through continual practice in moving the fingers and toes, in kicking, creeping, and raising itself, in crowing and babbling, wins the mastery over his organs; the boy that romps with others, and "can no more help running after another boy who runs provokingly near him than a kitten can help running after a rolling ball *—all these do not imitate serious action, whose organ has been dormant for an interval "longer than ordinary," but rather, impelled by irresistible impulse, they make their first preparations for such activities in this way.

Spencer's theory of play is therefore unsatisfactory, so far as concerns the adequacy of its explanation of the problem by means of the principle of imitation of previously accomplished serious activities of the individual.† And since in all the cases cited there is really no imitation of other individuals—that is, no "dramatization of the acts of adults" of which Spencer elsewhere treats—it appears that this principle of imitation can not be taken as a universal explanation of play. Nor can I agree with Professor Wundt when he says, in his Lectures on Human and Animal Psychology: "We regard certain actions of the higher animals as play when they appear to be imitations of voluntary acts. But they can be recognised as imitations because the result striven for only appears to be such, while the real end is the production of certain pleasurable effects, which are connected, though as mere accompaniments only, with real voluntary action This is as much as to say that animal play is in general terms identical with that of the human being. For this is, at least in its simpler form, and especially as it appears in the play of children, 'imitation of the business of practical life' stripped of its original aim and having a pleasurable mental effect." ‡ Wundt, in his

* W. James, The Principles of Psychology, London, 1891, II, p 427.

† See also the fine passage in Von Hartmann, Philos. d. Unbewussten, 10 Aufl., i, p. 179 f.

‡ W. Wundt, Vorlesungen über die Menschen und Thierseele, 2. Aufl., 1892, p. 388 (Eng. trans., p. 357).

Ethik, in which he is evidently influenced by Spencer, sets forth this idea perhaps more clearly. "Play," he says there, "is the child of work. There is no kind of play that does not find its prototype in some form of serious activity which naturally precedes it." *

It is, of course, undeniable that many plays originate in such imitation, but a glance over the passages cited above is sufficient to show that the most important and elementary kinds of play can be attributed neither to imitative repetition of the individual's former acts, nor to imitation of the performances of others.

If, then, Spencer's theory becomes so far untenable through the deviation of its imitative principle from Schiller's, our next step is evidently to inquire whether Schiller's idea alone would be satisfactory. Can it be admitted that accumulated superabundance of energy alone suffices to explain all the phenomena of play in animals? In order to get at the full meaning of this conception we must consider the psychological aspect of the surplus energy theory, as well as its merely physiological side. No doubt superabundant physical activity may often be considered as the psychological expression of exuberant spirits. This comes very near the idea that the play of animals and human beings originates in such physically conditioned dispositions; it is only necessary to instance the great influence good weather and comfortable temperature have on animals and men. Karl Müller notices this in an article on The Mental Life of Higher Animals in connection with the great influence the weather has on bird-songs, and says further: "Does this belong to the sexual instinct? Or has not rather the sense of comfort and well-being the

* Ethik, 1886, p. 145.

most influential part in it? Look at the healthy boy as he runs outdoors with his bread and butter. We often see him break forth into the most childish expressions of delight over the joy-bringer in his hand, and this delight in the thought of eating will show itself in leaping and running, and often in singing, the more excessively the more feeling or temperament controls him. And though the more advanced adult may not express his pleasurable excitement in singing, he does by whistling." * Th. Ziegler makes use of the same idea: "Joy in life, consciousness of strength, and the feeling of power—in short, the feeling of pleasure as such, is in its primitive and original meaning the beginning and the end of play for children." † And W. H. Hudson says in his wonderful book, The Naturalist in La Plata: ‡ "My experience is that mammals and birds, with few exceptions—probably there are really no exceptions—possess the habit of indulging frequently in more or less regular or set performances, with or without sound, or composed of sound exclusively, and that these performances, which in many animals are only discordant cries and choruses, and uncouth, irregular motions, in the more aerial, graceful and melodious kinds take immeasurably higher, more complex, and more beautiful forms." # "We see that the inferior animals, when the conditions of life are favourable, are subject to periodical fits of gladness, affecting them powerfully, and standing out in vivid contrast to their ordinary temper. And we

* Westermann's Illustrirte Monatshefte, 1880, pp. 239, 240.

† Th. Ziegler, Das Gefühl, 1893, p. 236

‡ First edition. London: Chapman and Hall, 1892. Third edition, 1895 (my citation is from this). See the brilliant criticism of the work by Wallace, in Nature, April 14, 1892.

Loc. cit., 264.

know what this feeling is—this periodic intense elation which even civilized man occasionally experiences when in perfect health, more especially when young. There are moments when he is mad with joy, when he can not keep still, when his impulse is to sing and shout aloud and laugh at nothing, to run and leap and exert himself in some extravagant way. Among the heavier mammalians the feeling is manifested in loud noises, bellowings, and screamings, and in lumbering, uncouth motions—throwing up of heels, pretended panics, and ponderous mock battles. In smaller and livelier animals, with greater celerity and certitude in their motions, the feeling shows itself in more regular and often in more complex movements. Thus *Felidæ*, when young, and very agile, sprightly species like the puma, throughout life simulate all the actions of an animal hunting its prey. . . . Birds are more subject to this universal joyous instinct than mammals, more buoyant and graceful in action, more loquacious, and have voices so much finer, their gladness shows itself in a greater variety of ways, with more regular and beautiful motions, and with melody." *

There is certainly no question that from the conception of physical and mental overflow of energy as it is laid before us in this series of pictures, a knowledge of one of the most important characteristics of the play condition is obtained. The physiological impulse that impels the latent powers to activity, and that mental joyousness whose highest point of development Schiller has justly recognised as the feeling of liberty, certainly form one of the most obvious characteristics of play. But it is quite as certain that the question whether by

* *Loc. cit.*, 280 f.

it a full comprehension of human and animal play can be obtained, must receive a negative answer; for, while simple overflow of energy explains quite well that the individual who finds himself in a condition of overflowing energy is ready to do something, it does not explain how it happens that all the individuals of a species manifest exactly the specific kind of play expression which prevails with their own species, but differs from every other. "Every species," says Hudson most truly,* "or group of species has its own inherited form or style of performance; and however rude and irregular this may be, as in the case of the pretended stampedes and fights of wild cattle, that is the form in which the feeling will always be expressed." Such a fact, depending as it does on the phenomena of hereditary transmission, evidently can not be explained by simple overflow of energy in an individual. Spencer has attempted to make use of the theory of imitation to point out the how and why of play activity. But we have seen that the most elementary and important plays can not be referred to it. It thus becomes necessary to call in the aid of some other conception of the subject. The solution of the problem is near at hand. Instead of pressing the idea of imitation exclusively, it is necessary to include that of instinct in general. Spencer himself has approached the right understanding of the problem. When he asks, What acts are chiefly imitated?—he reaches the conclusion, chiefly such actions as "in the life of this particular creature play the most important rôle." †

And proceeding to give some examples of this, he

---

* *Loc. cit.*, 281.

† Principles of Psychology, ii., p. 709. See Wallaschek as above: the imitation of actions that are "essential for the preservation of life."

points out that these important activities are instincts, in particular destructive and robbing instincts. Thus it is only necessary for him to modify his theory of imitation to stand directly in the presence of the right conception of play which lies so near his own. What form would the theory of play take in this case? Something like this: The activity of all living beings is in the highest degree influenced by hereditary instincts—that is, the way an animal of a particular species controls his members and uses his voice, the way he moves about in his natural element, supplies himself with food, fights with other animals, or avoids them—his manner of doing all these things is governed fundamentally by inherited instincts. When, now, there is on the one hand little demand for the serious activity of such instincts, and, on the other hand, the reintegration of nerve energy so far surpasses its expenditure that the organism requires some discharge of the accumulated supply of force—and both conditions are likely to be the case in youth *—then such instincts find expression even without serious occasion. The kitten treats a scrap of paper as its prey, the young bear wrestles with his brother, the dog which after long confinement is set free hunts aimlessly about, etc. But such actions are exactly what we mean by the word play.†

Paul Souriau seems to occupy a position similar to this in an interesting article ‡ where he advances the following idea: There are various grounds for the pleas-

* Also with animals in confinement. Spencer has specially alluded to this.

† Thus the imitative impulse appears as a special instinct related to the others. Concerning its significance I shall speak later.

‡ Le plaisir du mouvement, Revue Scientifique, iii série, tome xvii, p. 365 ff. L'esthétique du mouvement, Paris, 1889, p. 11 ff.

ure that animals almost universally take in movement. One of them is found in the fact that the animal is obliged to have a great capacity for movement in all the tasks of its life, for obtaining food, fleeing from its enemies, etc., and accordingly is endowed by nature with a correspondingly great feeling of the necessity for movement. When there is no occasion to give free play to this feeling, of necessity the confined impulses seek to break through all restrictions, even without serious motive, and so play arises. "Hence the movements of captive animals, of the lion who walks up and down his cage, of the canary bird that hops from perch to perch." So the necessity for movement controls even an inactive existence. For Souriau, too, there are inherited instincts that lead to play when superfluous nervous energy is present and the occasion for serious activity wanting.*

Such a conception as this, which does not need the principle of imitation, seems to me to be much nearer the truth. If we glance backward from this point of our inquiry we perceive that the essential points of the whole question have shifted considerably. At first the idea of the overflow of energy stood predominantly in the very centre of our mental horizon. But soon it appeared that for a full estimate of play it was necessary to consider something else. Now that we have found this something else to be instinct, the principle of surplus energy begins to lose some of its original importance. For it is now apparent that the real essence of

* G. H. Schneider expresses a similar view. He, too, places instinct more in the foreground, but without recognising the fact that the chief significance of the Spencerian principle would thus be imperilled. Der thierische Wille, 1880, p. 68. Der menschliche Wille, 1882, p. 201 f.

play, the source from which it springs, is to be sought in instinct. It is an essential fact that the instincts are constantly lurking in ambush ready to spring out on the first occasion. A condition of surplus energy still appears as the *conditio sine qua non*, that permits the force of the instincts to be so augmented that finally, when a real occasion for their use is wanting, they form their own motive, and so permit indulgence in merely sportive acts. Here I reach the limits of a merely physiological explanation of play. But before going a step further in the criticism of the overflow-of-energy theory by seeking to find a standpoint which includes the biological significance of play, I may here consider another theory which at first appears to be diametrically opposed to that of surplus energy. I mean the conception which obtains, especially in Germany, that play is for recreation. Steinthal has recently shown * very beautifully how recreation may be considered from its intrinsic significance to mean making one's self over—that is, creating anew, restoring lost powers, both physical and mental Such restoration can be had partially by means of sleep and nourishment. But in recreative play strength is needed to win strength. This idea is advanced by many. Guts Muths entitles his collection of games, Games for the Exercise and Recreation of Body and Mind.†

Schaller says that to the cultivated consciousness play presents itself somewhat as follows: An occupation not directed to the satisfaction of simply natural requirements or to the discharge of the practical busi-

---

* H Steinthal, Zu Bibel und Religionsphilosophie. Vortrage und Abhandlungen, new series, Berlin, 1895, p. 249.

† First edition, 1793; eighth edition, 1893.

ness of life, but securing rather the end of recreation.* Lazarus directs us, when we need restoration, to flee from empty idleness to active recreation in play.† The Jesuit Julius Cæsar Bulengerus begins his book on the games of the ancients with these words: "Neque homines neque bruta in perpetua corporis et animi contentione esse possunt non magis quam fides in cithara aut nervus in arcu. Ideo ludo egent. Ludunt inter se catuli equulei, leunculi, ludunt in aquis pisces, ludunt homines labore fracti, et aliquid remittunt, ut animos reficiant." ‡ But the most attractive exposition of the theory of recreation is given in an old legend quoted by Guts Muths.# John the Evangelist was once playing with a partridge, which he stroked with his hand. A man came along, in appearance a sportsman, and beheld the evangelist with astonishment because he took pleasure in a little creature which was of no account. "Art thou, then, really the evangelist whom everybody reads and whose fame has brought me here? How does such vanity comport with thy reputation?" "Good friend," replied the gentle John, "what is that I see in your hand?" "A bow," answered the stranger. "And why do you not have it always strung and ready for use?" "That would not do. If I kept it strung it would grow lax, and be good for nothing." "Then," said John, "do not wonder at what you see me do"

Here, then, there seems to be an irreconcilable conflict. The Schiller-Spencer theory allows the accumulated surplus of energy to expend itself in play; the rec-

* J. Schaller, Das Spiel und die Spiele, Weimar, 1861.

† M. Lazarus, Ueber die Reize des Spiels, Berlin, 1883, p. 48 ff.

‡ De Ludis privatis ac domesticis Veterum, 1627, p. 1.

Guts Muths, *loc. cit*, 22 f.

reation theory, on the contrary, finds in the very acts restoration of the powers that are approaching exhaustion. There they are wastefully cast off, here, thriftily stored away. Is it not remarkable that the same object can present itself to the observer in ways so contrary? A closer examination, however, shows that in this case the contradiction is only apparent. In fact, the two ideas can in many cases be so developed that they appear as different aspects of the same conception mutually explanatory of each other. When, for example, a student goes to have a game of ninepins in the evening, he thus tones up his relaxed mental powers at the same time that he finds a means of relieving his accumulated motive impulses, repressed during his work at the desk. So it is the same act that on the one hand disposes of his superfluous energy, and on the other, restores his lost powers. This is true in all cases when play can be considered as recreation. The recreation theory is thus, so far as it has any value at all, not contradictory, but rather supplementary to the Schiller-Spencer idea of play. An exhaustive criticism of the recreation theory, in so far as it claims to explain play, I do not consider necessary in a book treating of the play of animals. For it must be evident to any one, on reflection, that this idea, which may be very effective in a limited sphere, could not be justifiably expanded for application to the whole field of play. It occupies too much the standpoint of the adult who seeks recreation in a "little game" after the burden and heat of the day. That play can furnish recreation is not questioned, only that the necessity for recreation originates play. That the young dog romps with his fellows because he feels the need of recreation no one will seriously affirm. Evidently the advocates of the recreation theory, as a rule, know very little about the play of animals, and probably they have no conception of the extent of the subject. But the child whose whole mental life, as J. Schaller rightly remarks,* partakes predominantly of the character of play, must bear witness to the fact that while play may satisfy in many cases the need for recreation, it most certainly does not originate in it. I have been obliged to give special attention to the recreation theory, because it seemed to contradict the doctrine of surplus energy. It has now been shown that this is not the case. In seeking to go a step further in my criticism of the Spencerian theory, I find no support in the recreation idea, but must attempt to go on independently. Let us present clearly to our minds the position of our inquiry. Setting out with the overflow-of-energy idea, we found that Spencer's connection of this principle with that of imitation was not applicable to all play. Thus the expectation of explaining it all by means of surplus energy alone was found to be untenable. We then went on to include the idea of instinct. The overflow of accumulated vigour no longer appeared as the source of play, but yet as its *conditio sine qua non.* As now I proceed in the following pages to throw doubt also upon this formulation of the Schiller-Spencerian principle, I wish to avoid misunderstanding by making it clear at the outset that I do not underestimate the worth of that idea. It only seems to me that, even considering it as a mere *conditio sine qua non* of play, there is still a large territory to be accounted for outside of its limits. However, the overflow of energy is sufficiently important, and must be considered still the most favourable though not the necessary condition of play.

---

* Das Spiel und die Spiele, 1861, p. 2.

Going on now to the arguments that ground my own opinion, it can very easily be shown that the facts do not point to the universal or essential value of the Schiller-Spencerian principle. Certainly in innumerable cases the superfluity of unemployed energy gives an impulse to play, but in many others one is impressed with the fact that instinct is a power in itself which does not need special accumulated stores of energy to bring it into activity. Some examples will make this clear. Notice a kitten when a piece of paper blows past. Will not any observer confirm the statement that just as an old cat must be tired to death or else already filled to satiety if it does not try to seize a mouse running near it, so will the kitten, too, spring after the moving object, even if it has been exercising for hours and its superfluous energies are entirely disposed of? Or observe the play of young dogs when two of them have raced about the garden until they are obliged to stop from sheer fatigue, and they lie on the ground panting, with tongues hanging out. Now one of them gets up, glances at his companion, and the irresistible power of his innate longing for the fray seizes him again. He approaches the other, sniffs lazily about him, and, though he is evidently only half inclined to obey the powerful impulse, attempts to seize his leg. The one provoked yawns, and in a slow, tired kind of way puts himself on the defensive; but gradually instinct conquers fatigue in him too, and in a few minutes both are tearing madly about in furious rivalry until the want of breath puts an end to the game. And so it goes on with endless repetition, until we get the impression that the dog waits only long enough to collect the needed strength, not till superfluous vigour urges him to activity. I have often noticed that a young dog whom I have taken for a long walk, and who at last, evidently tired out, trotted behind me in a spiritless manner very different from his usual behaviour, as soon as he was in the garden and spied a piece of wood, sprang after it with great bounds and began playing with it. Just so we see children out walking who are so tired with their constant running about that they can only be kept from tears by coaxing, yet quickly set their tired little legs in motion again and deny their fatigue if an opportunity offers for play. Of children and young animals it is true that, except when they are eating, they play all day, till at night, tired out with play, they sink to sleep. Even sick children play, but only to the extent that their strength admits of it, and not as it exists overabundantly. Similar observations may be made with regard to the playing adult in many cases A student who has worked all day with a mental strain, so that he can hardly collect his thoughts for any serious effort, sits down in the evening to the mock battle of a card table and takes his part in the game with spirit for its complicated problems. "If any one will analyze the mental operations belonging to a single game of cards, the chains of reasoning which each player carries on for himself and attributes to the others, in order to plan for circumventing them, he will be much surprised at the variety and inexhaustible richness of mental activity displayed." * Can we speak in such cases of a superfluity of mental energy that originated from the fact of longer rest than usual?

A soldier or a banker who is engaged day by day in an exciting struggle with the caprices of fortune hurries to the gaming table, and for half the night, wavering between hope and fear, strives to produce the same sen-

* Lazarus, Reize des Spiels, p. 116.

sations. Must it not be admitted that he does not play for recreation nor for the relief of stored-up energy? It is the simple force of the demon instinct that urges and even compels to activity not only if and so long as the vessel overflows (to use a figure of speech), but even when there is but a last drop left in it. The theory of overflowing energy requires that first and necessarily there shall be abounding vigour; from it the impulse must originate. Superabundant life compels itself to act, says Spencer. The instincts would in that case be only the bed prepared for the self-originated stream to flow in. I maintain, on the other hand, that though this often appears true, it does not always prove to be so.*

It is not necessarily true that the impulse results from the overreadiness and straining of the nervous system for discharge. Notice the kitten that lies there lazily, perhaps even softly dozing, till a ball rolls toward it. Here the impulse comes from an external excitement that wakes the hunting instinct. If the kitten has a particular need for motor discharge she will play of

* Even in the case where the Spencerian theory appears to be most satisfactory—namely, that of the playful acts of animals in confinement, the monotonous walking up and down in the cage, the gnawing and licking the woodwork—is primarily not an instance of overflowing energy, but rather of thwarted instinct. Thus Lloyd Morgan says: "The animal prevented from performing his instinctive activities is often apparently unquiet, uneasy, and distressed. Hence I said that the animals in our zoölogical gardens, even if born and reared in captivity, may exhibit a craving for freedom and a yearning to perform their instinctive activities. This craving may be regarded as a blind and vague impulse, prompting the animal to perform those activities which are for its own good and for the good of the race to which it belongs." Animal Life and Intelligence, 1891, p. 430.

course. But when this need is not present, as is the case in our example, she still leaps after the ball; and only when disabled through utter fatigue would the cat fail to obey the impulse. The physiological conditions that lead a young animal to play at hunting need not be any other than those which enable an adult animal to pursue its natural prey.

If, therefore, these facts lead us to expect to find the chief problems of play in our conception of instinct, they also force upon us a consideration of the great biological significance of play. For even if I should not succeed in convincing the reader that superabundance of nerve energy is not even a *conditio sine qua non*, but rather only a particularly favourable condition for play, I have still every right to maintain that the Schiller-Spencer theory is unsatisfactory; for while it attempts, it is true, to make clear the physiological conditions of play, this theory has nothing to say about its great biological significance. According to it, play would be only an accidental accompaniment of organic development. For the advance toward perfection, due to the struggle for existence, brings it about that the more highly developed animals have less to do than their powers are competent for. Opposed to this view is the very general conviction among those who study animals that the play of young animals especially has a clearly defined biological end—namely, the preparation of the animal for its particular life activities. I have heard this explanation of play given in similar terms by foresters and by zoölogical specialists. Thus Paul Souriau says, in the article already referred to: "The necessity for movement is especially great in youth, because the young animal must try all the movements that he has to make later, and also exercise his muscles and joints to develop them. We know that all animals have a tendency to make use of a certain amount of energy, determined not by the accidental needs of the individual, but by the needs of the species in general." But if this is the case, play itself is not merely a result of the accidental needs of the individual, but rather an effect of natural selection, which works for anything that is serviceable for the preservation of the species. The observation of the different kinds of play is sufficient to establish this. Most plays of young animals—and it is this that must always present the essential problem in a theory of play—act for the preservation of the individual, all for the preservation of the species. At the same time the natural—that is, the self-originated—plays of human beings are to be considered as practice that is useful not only to the individual, but also to the race. "*Pro patria est, dum ludere videmur*" is the motto that Guts Muths has placed in the front of his book.

Can a phenomenon that is of so great, so incalculable value possibly be simply a convenient method of dissipating superfluous accumulations of energy? In all this there seems nothing to hinder the assumption that the instincts operative in play, like so many phenomena of heredity, first appear when the animal really needs them. Where, then, would be the play of the young? It would not be provoked either by overflowing nervous energy or by the need for recreation. Yet the early appearance of this instinct is of inestimable importance. Without it the adult animal would be but poorly equipped for the tasks of his life. He would have far less than the requisite amount of practice in running and leaping, in springing on his prey, in seizing and strangling the victim, in fleeing from his enemies, in fighting his opponents, etc. The muscular system would not be sufficiently developed and trained for all these tasks. Moreover, much would be wanting in the structure of his skeleton, much that must be supplied by functional adaptation during the life of each individual, even in the period of growth. The thought presents itself here that it must be the iron hand of natural selection that brings into bold relief without too compelling insistence and apparently without serious motive—namely, by means of play—what will later be so necessary. There need not be any particular superfluity of energy; so long as only a small remnant of unemployed force is present the animal will follow the law that heredity has stamped upon him.

Thus we see that the explanation of play by means of the overflow-of-energy theory proves to be unsatisfactory. A condition of superabundant nervous force is always, I must again emphatically reiterate, a favourable one for play, but it is not its motive cause, nor, as I believe, a necessary condition of its existence. Instinct alone is the real foundation of it. Foundation, I say, because all play is not purely instinctive activity. On the contrary, the higher we ascend in the scale of existence the richer and finer become the psychological phenomena that supplement the mere natural impulse, ennobling it, elevating it, and tending to conceal it under added details.

But the fundamental idea from which we must proceed is instinct. My first task must be the examination of instinct; and after a longer, but I hope not altogether uninteresting, exposition, I shall return to the points made above and give them more adequate treatment.

CHAPTER II.

PLAY AND INSTINCT.

WOULD it not be building on water or shifting sand to attempt the explanation of a psychological phenomenon by means of the mere concept of instinct? "The word instinct," remarked Hermann Samuel Reimarus in 1760, "has been until now so vague and unsettled that it scarcely had any certain meaning, or rather it had the most various uses." * This was still quite true up to the middle of the present century of the topic as a whole, and it will probably always continue to be true in regard to many details. "In speaking on instinct," says Ribot, with laconic brevity, "the first difficulty is to define it." † Since the time of Darwin, however, a great and important forward step has been taken, and Darwinism has assumed of late years a form that offers a fixed point of departure for the investigation of the problem that concerns us in this chapter. It is by no means my intention in what follows to give a history of the idea of instinct—a task never yet undertaken, to my knowledge. Still, it is necessary to clear the way for the comprehension of the problem and the appreciation of the view which I shall advocate, by

---

* H. S. Reimarus, Allgemeine Betrachtungen über die Triebe der Thiere, hauptsächlich über ihre Kunsttriebe, Hamburg, 1773.

† Th Ribot, L'Hérédité psychologique, Paris, 1894, p. 15.

a glance at the most important positions of modern thought. The following points of view may be distinguished:

1. The transcendental-teleological conception: (*a*) the theological, (*b*) the metaphysical, explanation of instinct.

2. The point of view which repudiates the notion of instinct.

3. The Darwinian solution, by means of (*a*) the transmission of both acquired and congenital characters; (*b*) the transmission of acquired characters only; (*c*) the transmission of congenital characters only.

Very early in modern thought we see the theological form of the transcendental-teleological conception of instinct brought forward by Descartes. For while he, following the Spaniard Pereira, denied to animals a reasoning intelligence, and considered them as mere machines or automata, he advocated the idea that the apparent intelligent actions of animals are to be traced directly to divine influence. The almost marvellous suiting of means to end seen in the actions of many animals, especially those displaying constructive instincts, furnishes sufficient ground for a similar opinion among many not at all inclined to deny all intellectual life to animals. (The strict Cartesian doctrine was for a long time so influential that the celebrated Leroy, through fear of persecution by the Sorbonne, published his letters on animal intelligence * as the work of a "physician of Nuremberg.")

The idea that these mysterious instinctive capabilities are directly implanted in the animal by God had a

* Ch G. Leroy. Lettres philosophiques sur l'intelligence et la perfectibilité des animaux, 1764, new edition, 1802.

great attractive power for religious natures, and especially so at the epoch of the Enlightenment, that period of reflective thought when the favourite attitude was one of "adoring contemplation" of the Creator's power. A naive conception of the universe like that of Gellert, for instance, who informs us in one of his poems that God called the sun and moon into existence for the purpose of dividing the seasons, naturally impels its holders to similar conclusions with regard to the adaptation of animal instincts. Two examples from this period and three modern ones may serve to illustrate this conception. Romanes quotes this remark of Addison's: "I look upon instinct as upon the principle of gravitation in bodies, which is not to be explained by any known qualities inherent in the bodies themselves, nor from any laws of mechanism, but as an immediate impression from the first mover and divine energy acting in the creatures." *

Reimarus regards instinct as a direct proof of the existence of God. His work, referred to above, contains a chapter on knowledge of the Creator through animal art-impulses, in which he expresses the opinion that such powers of body and soul as animal instincts disclose surpass the forces of Nature, showing us the "wise and good Author of Nature who has appointed for every animal the powers necessary for his life."

A definition from the eighth edition of the Encyclopædia Britannica may be mentioned as a modern example: † "It thus remains for us to regard instinct as a mental faculty, *sui generis*, the gift of God to the lower animals, that man in his own person and by them might be relieved from the meanest drudgery of Nature."

* G. J. Romanes, Animal Intelligence, p. 11.

† Cited from Romanes's Darwin and after Darwin, i, p. 290.

Brehm mentions a professor of zoölogy * by whom the old theory of instinct was set forth in its crude dualistic form—which was at that time combated most energetically by the opposers of the word instinct—i. e., that animals have only instinct and no reasoning powers, while man has reasoning powers and no instinct. "We know well," says this zoölogist, "that a being capable of adapting means to his ends must be a reflecting, reasoning being, and that in this world man is the only such being. An animal does not think, does not reason, nor set itself aims, and therefore, if it acts intelligently, some other being must have thought for it. A higher law provides the ways and means of its defence. The acts of men alone are governed by their own reason. Deep thought is doubtless disclosed in the actions of animals, but the animals did not think them any more than does a machine whose work represents an embodied chain of reasoning. The bird sings entirely without his own co-operation; he must sing when the time comes, and he can not do otherwise, nor can he sing at any other time. The bird fights because fight he must by order of a higher power. The fact is evident that the animal does not consciously fight for any special thing, such as the undisturbed possession of the female, nor seek by his struggles and effort to attain it. He acts as a mere creature of Nature under her stringent laws. It is not the animal that acts, for he is impelled by a higher power to altogether fixed courses of conduct. Parent birds can not deviate from a certain fixed method of rearing their young; both must work and help in the process; a command from above compels them to stay and work

* V. B. Altum, Der Vogel und sein Leben, Münster, 1875, fifth edition, p. 6 f., 114, 126; 13 f., 138, 141.

together. This is all that a happy marriage means to birds. There is no freedom, no voluntary action, no play of varying moods, no life of emotion or of thought to be expressed in the animal's actions. Without knowing what he does or why he does it, he makes directly for Nature's goal." *

The well-known zoölogist Wasman refers instinct to the Supreme Power, but with greater moderation. He holds that in instinctive acts themselves feeling and presentation may be present, but, so far as instincts are not explicable by the animal's own intelligence, he refers them to the Creator's influence. "Since," he says, "animals do not know the end of their instinctive actions, so much the less can they consciously pursue it. There must be a higher intelligence present, which not only knows the end but has ordered it. This intelligence can be no other than that of the Creator who has arranged the order of Nature, and made everything conducive to the proper preservation of that order. The adaptability of the several instincts of unreflecting brutes, as well as their correlation to those of other members of creation, must have its origin in creative intelligence." †

The efforts of metaphysicians to find a solution of the question are of a similar character. The spiritual principle is, of course, substituted for the Christian's God, but the transcendental-teleological view is retained. A few citations may be useful here too. Schel-

* A. E. Brehm, Thierleben, vol. i., p. 21.

† E. Wasman, Die Zusammengesetzten und Gemischten Kolonien der Ameisen. Münster, 1891, p. 214. See, also, the interesting article by O. Flügel, Zur Psychologie und Entwickelungsgeschichte der Ameisen, Zeitschrift für exacte Philosophie, vol. xx, p. 66.

ling puts All-pervading Reason in the place of a personal God. "Animals," he says, "in their acts express or witness to the All-pervading Reason, without themselves reasoning. Reason is in what they do without being in themselves. They may be said to reason through the force of Nature, for Nature is reason." And likening instinct to gravity, as Addison did, he reaches the conclusion that "the animal is held by instinct to the absolute Substance as to the ground by gravitation." *

G. F. Schuberth derives instinct from the "world-soul." † K. C. Carus says it is "the unconscious working of the Idea" that produces organic adaptation and beauty, and also instinctive activity ‡ Similar to this is E. von Hartmann's tracing of instinct to the "Unconscious." An exact student of Darwinian literature, he recognises the Darwinian principles only as means or instruments used by the Unconscious, in which alone the ultimate explanation is to be sought.#

So much for the transcendental-teleological theory. I am far from concurring with the many modern investigators who regard all religious or metaphysical ideas with contempt, seeing in the former a disease of youth, and in the latter youthful sentimentalism not worthy of the serious consideration of riper years.△ In a dec-

---

* System der gesammten Philosophie und der Naturphilosophie insbesondere, p. 238.

† Allgemeine Thierseelenkunde, Leipsic, 1863, p. 14, 22.

‡ Vergleichende Psychologie. Vienna, 1866. p. 59.

Das Unbewusste vom Standpunkt der Physiologie und Descendenztheorie, in the third volume of his Philosophie des Unbewussten, 1889, p. 271.

△ This false idea is referable to the principles of A. Comte. who laid down the "fundamental law" of the three stages of development in man—the theological or fiction stage, the metaphysical or abstract stage, and the scientific or positive stage—and likened

ade when we again stand at a turning point of time, when in polite literature the word is, Naturalism is dead; when plastic art turns toward a manifold, mystic, new idealism; when a Neo-vitalism is arising in biology; when a Brunetière proclaims with cool audacity the bankruptcy of positive science *—such a time is hardly a suitable one for the too confident assertion of the all-sufficiency of the exact sciences. Weismann has indeed given the title "The All-sufficiency of Natural Selection" to one of his essays, but in another of his writings occurs a figure which I like better. In contrast to the common opinion which likens empirical knowledge to a building resting on sure foundations and rising from a firm basement safely to the highest story, he says of the exact sciences: "They all build from above, and not one of them has found a basement yet—not even physics." † This is indeed true. The metaphysical problems do not float above us far off in the clouds while we peacefully do our work on the firm, enduring earth, but they are rather beneath us, and our clear empirical knowledge rests on their mysterious depths like the sun-reflected sails of a ship on dark waves. So long as this is so, man can not satisfy himself with the "unknowable" and the "ignorabimus" of positivism, but will constantly seek to fathom these bafflingly mysterious depths on which he is borne along. In this book, however, no attempt is made at a metaphysical solution of instinct

---

them to the three stages of individual development—childhood, youth, and manhood. Cours de philosophie positive, second edition, 1852, vol. i, pp. 14, 17.

* F. Brunetière, La science et la religion, Paris, 1895.

† Die Bedeutung der sexuellen Fortpflanzung für die Selectionstheorie, Jena, 1886, p. 60.

or any other problem. Metaphysics, or the first science, as its original name signified, should rather be called the last science. It belongs to the end rather than to the beginning of an inquiry. So it will be at the close of my book on human play that I shall speak of the metaphysical side of my subject, if at all. For surely whether its use is justifiable or not, the time is past when it could be appealed to as a means of approaching a subject before empirical research was attempted. The *merely* metaphysical grounding of phenomena will never again suffice.

As a result of this empirical tendency we see a strong opposition to the transcendental-teleological view arise in the second half of our century. It assumes the form of a double criticism, a negative and a positive. The one wishes to eliminate the word instinct altogether wherever possible. The other gives to it a new meaning, no longer involving the supernatural.

The repudiation or rejection of the conception of instinct arises from the fact that the attempt is made to explain all instinctive acts as the result of individually acquired experience and reflection. Of the many who have adopted this view, I notice only the more modern.*

Turning first to the great work of Alexander Bain, The Senses and the Intellect, we find there a long chapter on instinct, but in it no mention is made of the actions which we are usually accustomed to speak of as instinctive. Only reflex movements, such as heart beats, breathing, coughing, sneezing, gestures, etc., are referred to. Bain's view of real instinct is first developed in the

---

* For older scholars holding this view, see Fr. Kirchner, Ueber die Thierseele, Halle. 1890, and L. Büchner, Aus dem Geistesleben der Thiere, third edition, Leipsic, 1880.

section on "Associations of Volition," * where he seeks to show that such instinctive acts are acquired by individuals rather than inherited. In his companion work, The Emotions and the Will, he teaches, too, that heredity only explains the simple movements that can be attributed to reflex action. The development of these to complicated instinctive acts, he says, depends on the individual performance of the animal.†

Alfred Russel Wallace was formerly another opponent of the idea of instinct. He differed from Bain in denying to the word its application to even simple reflex action. "It is sometimes absurdly stated that the newborn infant 'seeks the breast,' and this is held to be a wonderful proof of instinct. No doubt it would be if true, but, unfortunately for the theory, it is totally false, as every nurse and medical man can testify. Still, the child undoubtedly sucks without teaching, but this is one of those simple acts dependent upon organization which can not properly be termed instinct any more than can breathing or muscular movement." ‡

Wallace believed, moreover, as Bain does, that instinctive acts must be learned by each individual. This appears most clearly in his brilliant essay on The Philosophy of Birds' Nests.#

---

* A. Bain, The Senses and the Intellect, third edition, London, 1868, p. 409.

† A. Bain. The Emotions and the Will, third edition, London, 1880, p. 53. Bain records observations of a newborn lamb to show that the so-called instinctive capabilities were acquired by it. But this should be compared with Hudson's notice of wild sheep: he often saw these stand on their feet five seconds after birth, and when one minute old run after the mother. The Naturalist in La Plata, p. 109.

‡ A. R. Wallace, Contributions to the Theory of Natural Selection, p. 206.

Loc. cit., p. 211 f.

It is generally asserted that birds would build nests like all others of their kind even if they had never seen them. That would undeniably be instinctive. "This point, although so important to the question at issue, is always assumed without proof, or even against proof, for the known facts are opposed to it. Birds brought up from the egg in cages do not make the characteristic nest of their species, even though the proper materials are supplied them, and often make no nest at all, but rudely heap together a quantity of materials." *

"With regard to the song of birds, moreover, which is thought to be equally instinctive, the experiment has been tried, and it is found that young birds never have the song peculiar to their species if they have not heard it, whereas they acquire very easily the song of almost any other bird with which they are associated." For Wallace, such cases are accounted for by simple imitation and in a slight degree by adaptation of the individual to new conditions.† However, he does not consider it impossible that the existence of pure instinct may be proved in some cases.‡

Later, Wallace changed his view entirely, and admitted inherited instinct. "Much of the mystery of instinct arises from the persistent refusal to recognise the agency of imitation, memory, observation, and reason as often forming part of it"; but with these elements depending on individual effort, he recognises the force of inheritance as one of the actual elements of

---

* *Loc. cit.*, p. 220. This shows the building tendency in spite of the abnormal conditions!

† His grounds bring to mind in part those of Condillac and Leroy.

‡ *Loc. cit.*, p. 230.

instinct. Indeed, he approaches Weismann's standpoint on this question, as I do.*

The discussion took on a more polemic form in Germany. Materialism made the attack. Carl Vogt, in the last chapter of his Pictures from Animal Life, speaks contemptuously of "so-called instinct." Brehm in his great work employs all the eloquence at his command against the "impossible doctrine of so-called instinct in animals." † And Büchner follows him with an exhaustive discussion. All these writers agree in attacking first the theological conception, to which their materialistic point of view is, of course, fundamentally opposed. And they naïvely assume that any other point of view is out of the question. Thus we find in Büchner this remarkable definition: "Men have fallen into strange ignorance and conceit in calling the unknowable soul-expression of animals instinct, a word derived from the Latin *instinguere* (to stimulate or incite), and therefore necessarily implying a supernatural stimulator or inciter." ‡ When the materialists become acquainted with Darwin's positive criticism of the old instinct idea, they agree indeed with it, but, passing by with slight notice his theory, they were not disturbed in their polemic against the "unfortunate word instinct." Buchner, especially, protests in several of his works sharply and persistently against the use of the word. He dwells on its variability of signification and on its mistaken employment, and considers parental teaching and individual experience and reflection the true sources of actions usually called instinctive. He points out after careful study that "the most of what was formerly ascribed to instinct

* A. R. Wallace, Darwinism, p. 442.

† Thierleben, second edition, i, p 20.

‡ L. Büchner, Kraft und Stoff, 1883, p. 471.

may be explained in ways altogether different and more natural, either as produced by genuine reflection and choice, or by experience, instruction, and information; or by practice and imitation; or by a particularly good development of the senses, especially of smell; or by custom and organization; or by reflex action, etc. For example, when the caterpillar uses the fibre which Nature provides for building its nest, for hanging itself from a tree and thus eluding the pursuit of its enemies; or when caterpillars shut up in a drawer eat off the paper lining and use it for making a cocoon; or when toads persist in devouring great quantities of ants which taste good but which they can not digest, although they know (?) that pain and illness must be the consequence; or when bees passionately consume honey mixed with brandy, which maddens and unfits them for work; when birds build their nests near human habitations for the purpose of using material such as thread and woollen scraps; or when, according to the observations of G. H. Schneider, certain crustaceans in captivity use bits of cloth and paper to hide under in the absence of weeds, though when both are present their choice is always in favour of the vegetable substance; or when bees, presented with a set of prepared cells, stop building cells and carry their honey to the finished ones; or when birds prefer an artificial nest box or an appropriated nest to the product of their own skill; or when ants seize strange nests in the same way and establish themselves comfortably there instead of building for themselves; or when many kinds of bees, instead of collecting their own honey, get a supply by robbing other hives; or when animals imitate the voice or the cries of other animals happening to be near for purposes of defence or enticement—in these and a thousand similar cases whose enumeration would fill a whole book, instinct can not be the cause or the occasion of a single one of such actions." *

The wealth of examples with which the author cleverly overwhelms us might be convincing to an uncritical reader. In truth, however, Büchner combats only the most extreme conception of instinct, that is hardly to be taken seriously in our day; just as in his antagonism to theology and metaphysics he attacks with his materialistic weapons only the extremest orthodoxy and the most abstruse speculation, but in such a manner that an unlearned reader might well get the impression that theology and metaphysics generally had received their death blow. What Büchner refutes is the idea of a direct and miraculous imparting by God to the animals of absolutely inflexible and inerrant instincts. It is surely possible to reject this view and yet believe in an instinct which acts under normal conditions suitably to ends, as inborn capacity in man and beast, without individual experience and without a knowledge of the end, but which may vary under different circumstances, and become in abnormal cases so unsuited to the supposed end that it may be said to "err." Moreover, Buchner and the other opponents of instinct can by no means claim that their idea is altogether contrary to the pre-Darwinian view, for the extreme instinct theory briefly outlined above was not by any means universally held even before Darwin. Thus Reimarus, who has been quoted already, and who was easily the most influential animal psychologist of his time, his General Observations on the Impulses of Animals pass-

* L. Büchner, Aus dem Geistesleben der Thiere, third edition, Leipsic, 1880, p. 16.

ing through many editions and being translated into French and Dutch—in this work (§ 98) Reimarus says: "The mechanical instincts of animals are not so fixed in every point that the creature is not left the power to modify them according to circumstances and the extent of his own knowledge." * And the first sentence of § 101 runs thus: "Animals may sometimes err in their impulses, though this seldom happens when they are entirely left to themselves."

The denial of inherited instinct can in no wise be regarded as established. Reimarus himself has controverted, on grounds which in essentials are not yet obsolete, those who regard instinct as an empty or meaningless word.† For example, he says in § 93: "Many mechanical instincts are practised from birth without experience, instruction, or example, and yet faultlessly. They are thus seen to be certainly inborn and inherited. . . . This is the case with all insect grubs that envelop themselves with a spun web, such as wasps and many caterpillars, bees, and ants. How can a worm that has lived scarcely a day, and that shut up in the dark ground or a little shell, possibly have acquired of itself such skill from experience or from lessons and examples?

* See the fragmentary posthumous publication, H. S. Reimarus's Angefangene Betrachtungen über die besonderen Arten der thierischen Kunsttriebe, Hamburg, 1773, introduction.

† I fully agree with A. Kussmaul in what he says of the "splendid work" of Reimarus, "which will stand for all time as a model for critical investigators in this subject" (Unters. über d. Seelenleben des neugeborenen Menschen, Leipsic, 1859, p. 5). The book of G. F. Meier (Versuch eines neuen Lehrgebäudes von den Seelen der Thiere, Halle, 1749) is also celebrated, but can not be compared with Reimarus' work, seeing that, excepting some observations on ants, it contains essentially only the typical "logical reasoning of the Enlightenment."

The same question may be asked about the animals that are hatched out by the sun on the sand, and as soon as they creep out of the egg hurry to the water without being shown the way; of the young duck, too, that in spite of the cries of the clucking hen betakes itself to the strange element. We have incontrovertible proof also that the mechanical impulses are innate and inherited in animals that are taken living from the mother's womb, and so could not by any possibility have seen any others or have learned to act as they do. The celebrated Swammerdam has made such an experiment with the water snail, which is born alive. He took a little one just ready for birth and placed it in water, where it immediately began to move about quite as well as the mother could. And this implies great skill, for, in order to sink, these snails retire into the shell and compress the air contained in the end-chambers, thus becoming heavier than the water. To rise, they come out a little, causing the inclosed air and their own body to occupy more space and so become lighter than the water. For surface swimming they turn over so that the shell is like a boat, the feet are extended on both sides, and an undulating movement like that of the land snail sends it slowly over the water. This skill and readiness in movement the snail cut from its mother's body has certainly not learned nor practised, but brought already developed into the world." I may here point out that Reimarus very rightly emphasizes the difficulty of learning an entirely new kind of movement. If, for example, sucking the breast "were not innate skill, so to speak, there is no reason why grown people should not do it as well as children, particularly as they are practised in various movements of the mouth, and even in sucking at other soft tubes. But, speaking for myself at least, I must own that I can no longer do it" * (§ 138).

It would hardly be necessary to cite further examples of inherited instinct † if the principle involved were not so vital to my purpose. As this is so, I give the views of two modern philosophers who both defend the idea of instinct, though from very different standpoints and without being in any special sense Darwinians.

E. von Hartmann gives the following among a great many other examples: "Caterpillars of the *Saturnia pavonia minor* eat the leaves of a shrub as soon as they emerge from the egg, go underneath the leaves when it rains, and change their skin from time to time; this is their whole existence, and in it not the least evidence of intelligence can be found. But the time comes for spinning their cocoon, and they build it firm and strong, with a double arch formed by gathering the fibres together at the top, so that they are very easy to open from the inside, but offer considerable resistance to any external force. Were this arrangement the result of conscious intelligence, a chain of reasoning something like this would be necessary: 'I am now approaching a chrysalis state, and, immovable as I am, I shall be exposed to attack; therefore I will inclose myself in a cocoon. Since I shall not be in a condition as a butterfly to effect an exit either through mechanical or through chemical means, as many other caterpillars do, I must therefore provide an opening. But at the same time,

* It may be mentioned that even the sceptic David Hume called instinct a primary gift of Nature, a degree of capability that ordinarily surpasses the animal's powers and can not be much bettered by the longest practice or use. An Inquiry concerning the Human Understanding, p. 99.

† See also A. and K. Müller, Wohnungen, Leben und Eigenthümlichkeit in der höheren Thierwelt, p. 8 f.

that this opening may not be used by my enemies, I must use such an arrangement of the fibrous web as will allow me to push out but will yet offer resistance to outside pressure, according to the principle of the arch.' This does seem too much to expect from the poor little caterpillar." *

Wundt cites the same example, originally suggested by Autenrieth,† as especially significant, and says, moreover: "If it were actually through a capacity for adapting means to an end that the bird produces its nest, the spider its web, and the bee its cell structure, a degree of intelligence would be required that man himself, in the course of a mere individual life, would hardly be capable of. A further proof of the fallacy of such an explanation is the regularity with which certain actions are performed by individuals of the same species where there is not always any association between them. Such association, of course, exists among the bee and ant tribes and among those animals whose young remain for some time after birth with their parents, but in numerous other cases the little creature begins its life independently. When the caterpillar creeps out of the egg its parents are long since dead, yet it prepares a cocoon like theirs. And, finally, there are many cases where the instinct-acts that seem to be intelligent appear to include a direct foresight of the future. How can this foresight possibly be intelligent when there has never been analogous experience in the individual's life? Nor has it received information in any way. When the moth incloses its eggs in a furry covering made of its own hair, the winter, which makes this warm wrapping necessary

* E. von Hartmann. Philosophie des Unbewussten. i. p. 79.

† J. H. F. Autenrieth, Ansichten uber Natur- und Seelenleben, 1836, p. 171.

for the preservation of the egg, has not yet come. The caterpillar has never experienced the metamorphosis for which it prepares." *

I conclude with a few examples from the numerous witnesses among modern scientists. "The instinct for flight to warmer lands," says Naumann,† "is born in migratory birds. Young ones taken from the nest and allowed to fly about freely in a large room sufficiently prove this. They circle about their prison at night, during their time of migration, just as the old birds do in confinement." Douglas Spalding experimented ‡ as Swammerdam also did. He took little chicks from the egg, put caps on them that covered the eyes until they were two days old. When these were removed one of them at once followed with its eyes and head a fly some twelve inches away. A few minutes later it picked at its own toes, and in the next moment sprang vigorously after a fly and devoured it. It ran at once, with evident assurance, to a hen brought near, and seemed to need no experience or association in all this to enable it to go over or around impediments, for these were its first lessons in life. Spalding has also shown experimentally that young swallows can fly without teaching as soon as they reach the proper age. Further he relates: "One day, after I had been stroking my dog, I reached my hand into a basket which held four blind kittens three days old. The smell of my hand made them spit and hiss in a ridiculous way." # Romanes succeeded in mak-

* It may be noted here that Lotze, too, holds the hypothesis of instinct as indispensable. See his larger Metaphysik. p 299

† J. A. Naumann, Naturgeschichte der Vogel Deutschlands, i, p 86.

‡ Macmillan's Magazine, February, 1873.

See Wesley Mills, The Psychic Development of Young Ani-

ing a quite similar experiment with young rabbits and ferrets.* Hudson once found some eggs of the *Parra jacana*. "While I was looking closely at one of the eggs," says he, "lying on the palm of my hand, all at once the cracked shell parted, and the young bird quickly leaped from my hand and fell into the water. I am quite sure that its sudden escape from the shell and from my hand was the result of a violent effort on its part to free itself, and it was doubtless stimulated to make the effort by the loud, persistent screaming of the parent birds, which it heard while in the shell. Stooping to pick it up to save it from perishing, I soon saw that my assistance was not required, for immediately on dropping into the water it put out its neck and with the body nearly submerged, like a wounded duck trying to escape observation, it swam rapidly across, and, escaping from the water, concealed itself in the grass, lying close and perfectly motionless like a young plover." †

Weinland reports of the snapping turtle: "For months these turtles emerge daily from eggs laid in the sand and moss, and it is noteworthy that the first movements of the little heads thrust out of the broken shell are those of snapping and biting." ‡

Preyer and Binet are firmly convinced that instinct is the source of the child's first attempts to walk. According to Binet's observations, children only a few weeks old really take measured steps when held up so that the soles of their feet rest on the floor.# In short,

---

mals, 1895, iv and vi, where Spalding's investigation is referred to as "somewhat overdone, though reliable in the main."

* G. J. Romanes, Mental Evolution in Animals, p. 164f. See also Hudson's Naturalist in La Plata, chap. vi, p. 89.

† The Naturalist in La Plata, p. 112.

‡ From Brehm's Thierleben, second edition, vii, p. 64.

A. Binet, Recherches sur les mouvements quelques jeunes

James is perfectly right when he says that the sitting hen, for example, needs no further experience or psychic process than the feeling that the egg is just "the-never-to-be-too-much-sat-upon object." *

From all this it appears that there is no doubt that inherited instincts exist, and that a positive rather than a negative criticism will be needed in dealing with this idea, which, indeed, is much easier to affirm than to explain. We at once reach the conclusion, however, that it is necessary to eliminate from the definition of instinct the transcendental-teleological method of conceiving it —a task which has been attempted by the promulgators of the Lamarck-Darwinian theory. "An important reason for the slow advance of scientific knowledge is the universal and almost unconquerable adherence to teleological conceptions, which are substituted for distinctly scientific ones. Nature may affect us ever so impressively and ever so variously, but it is all lost upon us because we look for nothing in her manifestations that we have not already read into them; because we will not permit her to make the advances, but are always trying with impatient, ambitious reasoning to approach her. Then, when in the course of centuries there comes one who draws near to her with a quiet, modest, and receptive mind, and lights upon innumerable phenomena which we, by our preoccupation, have overlooked, we are amazed indeed that so many eyes should not have seen them before in such clear light. This striving with unnecessary haste after harmony before the various tones which should compose it have been collected, this violent

enfants. [See also the experiments of Baldwin, Mental Development in the Child and the Race, second edition, chap. v, § 1.]

* W. James, The Principles of Psychology, London, 1891, vol. ii, p. 387.

usurpation by the intellectual powers of a realm where they have not undisputed sway, explain the unfruitfulness of so many thinkers for the advancement of science. It is difficult to say whether mere sentiment, which assumes no definite form, or much reasoning tending to no purport, has interfered more with our progress in knowledge." But who has written this masterly indictment of modern science? I fancy that one would not easily detect the author, did not certain artistic expressions betray him, and withal the perfect style which is as brilliant and as penetrating as a good sword.*

Lamarck published his theory of development in 1801, and extended it in the Philosophie zoologique, which appeared in 1809, as well as in the introduction to his work on the Histoire naturelle des animaux sans vertèbres. As the fundamental principle of his theory, he lays down the inheritance of acquired characters by individuals (especially functional adaptations). Darwin included this principle in his theory advanced in 1859, but perfected it by his more important and comprehensive conception of natural selection. According to it, not only functional adaptations, but also the inheritance of congenital characters produce changes in species, so that in each generation congenital "individual variations" appear of which the "fittest" always come out best in the struggle for existence, and are thus transmitted further (Spencer, Survival of the Fittest). In the whole organic world this principle rules, adapting means to ends without there being any end—that is, any conscious or voluntary end. The transcendental-teleological principle is thus excluded. "The

* Schiller, Ueber die ästhetische Erziehung des Menschen, thirteenth letter.

useful becomes the necessary as soon as it comes to be possible." *

Darwin himself attached greater importance to congenital qualities than to the inheritance of acquired ones, as clearly appears in his definition of instinct. He says in his Origin of Species: "It would be the most serious mistake to suppose that the greater number of instincts have been acquired by habit in one generation, and then transmitted by inheritance to successive generations. It can be clearly shown that the most wonderful instincts with which we are acquainted—namely, those of the hive-bee and of many ants—could not possibly have been thus acquired." And in the Descent of Man: "Some intelligent actions—as when birds on oceanic islands first learn to avoid man—after being performed during many generations become converted into instincts, and are inherited. . . . But the greater number of the more complex instincts appear to have been gained in a wholly different manner through the natural selection of variations of simpler instinctive actions." †

We see, then, that Darwin derives instinct from two distinct sources. The principal source is natural selection; the less important is the inheritance of intellectual capacity and then of acquired characters. Romanes follows him closely in his distinction between

---

* A. Weismann, Amphimixis, Jena, 1891, p. 159. [The author here includes a quotation from Kant's Physische Geographie (II. Th., Abs. i. § 3) showing that that philosopher had the idea of progressive development resulting from artificial selection, which in Darwin's mind led to the doctrine of Natural Selection. He refers also to Fischer's Geschichte d. neu. Philos., third edition, iii, p. 161.]

† The Descent of Man, chap. ii. See also a similar passage from Darwin's manuscript in Romanes's Mental Evolution in Animals, p. 209.

primary and secondary instinct. He says: "I shall allude to instincts which arise by way of natural selection, without the intervention of intelligence, as primary instincts, and to those which are formed by the lapsing of intelligence as secondary instincts." *

Romanes in turn has influenced some other animal psychologists. Thus Foveau de Courmelles says, in elaborating Romanes' distinction: "The primary instincts consist of non-intelligent habits devoid of adaptability, transmissible by heredity, themselves subject to variation and liable to become fixed. Secondary instincts are intelligent adaptations that have become automatic and hereditary." † And Lloyd Morgan, who refers to Romanes's treatment of the instinct idea as most masterly and admirable, likewise adopts the division into primary and secondary instincts, but, owing to the influence of Weismann and Galton, is very cautious about approaching the subject of the inheritance of acquired characters, and, consequently, that of secondary instincts. He accordingly attributes to these principles only a probable value. ‡

The great majority of modern animal psychologists, however, explain instinct by the Lamarckian principle of the inheritance of acquired characters alone, or almost alone. Their conception of instinct is something like this: Darwin had already pointed out its analogy to acts

* Mental Evolution in Animals, chap. xii I am well aware that the distinctions between "inherited" and "acquired," "primary" and "secondary," instincts are not exactly the same, but I can not go into these finer points here.

† Les facultés mentales des animaux, Paris, 1890, p. 55.

‡ Animal Life and Intelligence, p. 433. [The reader should turn to Lloyd Morgan's later work, Habit and Instinct, in which he gives up the "inheritance of acquired characters" altogether.]

in individual life which have become reflex through practice and repetition; the piano player reaches for the right key "mechanically," intuitively, though at first he could make the same movement only under the control of conscious will. In just the same way inherited instinct depends on a "lapsing of intelligence" (Lewes), but instead of being accomplished in a single life, it progresses in such a manner that the conscious practice of earlier generations becomes the reflex activity of later ones.* This is what is meant by the common designation of instinct as inherited habit or hereditary memory. I cite only a few examples: Preyer and Eimer use these expressions in defining instinct, and L. Wilser calls it hereditary skill or aptitude.† Wundt says, "Movements that originally appeared as simple or compound acts of the will, but later, either in the life of the individual or in the progress of race development, have become partially or entirely mechanical, we call instinctive acts." ‡ Th. Ribot,# with Lewes, calls instinct "*conscience éteinte*," and Schneider refers what he recognises as hereditary in instinctive acts to the practice and habit of ancestors. ‖ Thus he explains our instinctive fear in the dark as the inheritance of acquired association:

---

* This interesting passage is from Leroy: "What we regard as entirely mechanical in animals may be ancient habit perpetuated from generation to generation" Lettres philosophiques sur l'intelligence et la perfectibilité des animaux, new edition, Paris, 1802, p 107.

† W. Preyer, Die Seele des Kindes, Leipsic, 1890, p. 186. Eimer, Entstehung der Arten, i, p. 240. L. Wilser, Die Vererbung der geistigen Eigenschaften, Heidelberg, p. 9.

‡ W. Wundt, Vorlesungen uber die Menschen- und Thierseele, second edition. 1892. p. 422. [Eng. trans., p 388.]

L'Hérédité psychologique, fifth edition, p. 19.

‖ G. H. Schneider, Der thierische Wille, Leipsic, 1880, p. 146.

"Not only our savage ancestors but even those of later times who have not had the good fortune to live, as we of the present do, in circumstances rendered secure by orderly government, could not undertake the slightest journey, especially by night, with the carelessness with which we now in middle Europe tramp through the loneliest mountain pass or traverse the densest woods by day or night. They had much to fear from wild animals, especially bears, and from men, such as highwaymen and the famous robber knights, and in lonely woods and passes were never safe. Moreover, the feeling of fear which besets the young, especially when travelling alone on a dark night in a lonely wood or valley, is so universal that we are forced to connect it with the common experience of earlier generations, and consider it an inherited feeling." *

If this reference of instinct to the inheritance of acquired characters which we find is so general be correct, play can be explained about as follows: Our ancestors have throughout their whole lives made use of their arms and legs for every possible movement; accordingly, their descendants have in their earliest infancy the impulse to kick with the legs and to grasp everything in their hands. The forbears hunted animals; hence the hunt and chase games of the descendants. Our ancestors were obliged to hide from their enemies in a thousand ways; hence the hiding games of children. Thus Schneider says: "The boy does not now eat the sparrows, beetles, flies, and other insects that he eagerly seizes and perhaps tears to pieces, nor does he intend to devour the young birds that he takes from their nests in high trees, often at the peril of his

* G. H. Schneider, Der menschliche Wille, Berlin, 1882, p. 68.

life; but merely seeing these things wakes in him a strong impulse to plunder, hunt, and kill, apparently because his savage ancestors commonly gained their subsistence by such means. There is in him an intimate causal connection between the sight of certain free animals, or birds' eggs, and the impulse to plunder, slay, and rend. That this was the case with our animal ancestors we are convinced from the life of modern apes, which is sustained principally by means of spoil taken from smaller animals, especially insects, young birds, and birds' eggs" * "Girls, as well as boys, show in their play unmistakable signs of having inherited the characteristic habits of the race." † Thus play becomes the result of intelligent activity of preceding generations, a form of hereditary skill.

In the last decade, however, the general conception of instinct has undergone an essential transformation through August Weismann's neo-Darwinism. I can not here, of course, go thoroughly into the highly complex grounds of this theory of heredity.‡ Weismann postulates an hereditary substance carried on continuously through succeeding generations, the germ plasm (*Keimplasma* #) which is present in the so-called chromosomes, or colourless bodies of different shapes inside the cell nuclei ("chromatin bodies" or "chromatin nuclei"). He not only asserts in a general way that this substance inside the germ cells must have an exceedingly

* Der menschliche Wille, p. 62.

† Ibid., p. 32.

‡ See especially Die Continuität des Keimplasmas, Jena, 1885. Amphimixis oder die Vermischung der Individuen, Jena, 1891. Das Keimplasma, Jena, 1892.

Keimplasma, p. 32.

complicated structure* historically handed down, which, indeed, is undeniably the case, but in a more daring hypothesis he attempts to establish the essential elements of this structure: the molecules of germ plasm go in various ways to form Biophores, which determine the cell qualities; † these in turn form Determinants,‡ which again find their higher unity in the Ide, # these, again, are grouped in the Idant,‖ which is identical with the chromosome.

But this world of minute elements represented by the germ cells is, as I said before, continuous—that is, it is not produced anew in each individual, but persists with great stability throughout the countless successions of related life forms, building up organisms but never exhausted in the process, and not influenced by individual experience or by heredity. It may be figured as a creeping root, stretching far from the parent stock; single plants rise from it at different points, represented by the individuals of successive generations.^ If, then, a material so constituted is the only medium for the operation of heredity, there can be no transmission of acquired characters.

Weismann's theory taken as a whole is far from universally recognised as established. It has a great number of opponents, of whom I mention only Haeckel.◊

* Keimplasma, p. 82.

† Ibid., p. 53.

‡ Ibid., p. 76

Earlier called "Ahnenplasma" by Weismann, ibid., p. 84.

‖ Ibid., p 90

^ A. Weismann, Die Bedeutung der sexuellen Fortpflanzung fur die Selectionstheorie, Jena, 1886, p. 20.

◊ Haeckel, Naturliche Schöpfungsgeschichte, 1889, p 198. Anthropogenie oder Entwickelungsgeschichte des Menschen, 1891, preface.

Eimer,* Wilser,† Hertwig,‡ Romanes,# Herbert Spencer,|| Wundt,^ Sully,◊ and Ribot.‡ The truth is, it is not yet given to us entirely complete, for almost every work of the gifted author yet published shows some modification more or less important.‡ The weightiest point to be determined before the theory can be further developed is that of the relation of the individual to the hereditary substance or of the soma to the germ plasm. Does this germ plasm pervade the endless series of individuals with absolute continuity, changing only through its combination with that of other individuals (amphimixis)? Weismann formerly appeared to attribute absolute persistence to the germ plasm; indeed, he has, in one instance at least, emphasized this doctrine. Yet in 1886 he admitted that monads that are propagated by mere division may inherit acquired characters.‡ In 1891 he limits this possibility to unicellular structures without a nucleus.** In other directions, however, he has

* Th. Eimer, Die Entstehung der Arten auf Grund von Vererben erworbener Eigenschaften nach den Gesetzen organischen Wachsens, 1888.

† L. Wilser, Die Vererbung der geistigen Eigenschaften, 1892.

‡ O. Hertwig, Zeit- und Streitfragen der Biologie, vol. i. Präformation oder Epigenese?, 1894.

G. J. Romanes, Critical Examination of Weismannism.

|| Herbert Spencer, The Inadequacy of Natural Selection, 1893. A Rejoinder to Professor Weismann, 1893. Weismannism once more, 1894.

^ Wundt, Vorlesungen über die Menschen- und Thierseele, second edition, 1892, p 441.

◊ J. Sully, The Human Mind, 1892, i, p. 139.

‡ Th. Ribot, L'hérédité psychologique, 1894 preface.

‡ Since this was written the theory of "Germinal Selection" has been added to it.

‡ Die Bedeutung der sexuellen Fortpflanzung für die Selectionstheorie, Jena, 1886, p. 38.

** Amphimixis, Jena, 1891. (Weismann must of course hold to

weakened this position by the admission that the germ plasm may have only a "very great" but not absolute persistence. I do not refer to his granting the inheritance of diseases (these are, after all, only pollutions of the stream that may not essentially alter it), but to his admitting the possibility of modifying the germ plasm by changing nutriment and temperature.*

Next in order is his essay on External Influences as Aids to Development (1894), where he shows that he is not blind to the importance of external conditions. He here concedes that the development of germ plasm itself may be modified by means of changes in nutriment and temperature, while predispositions that remain latent under ordinary circumstances may be stimulated to activity by such "external aids." The fact that this is not the cause but only the occasion of the modification is especially emphasized, the cause being always the predisposition latent in the germ. That the persistent quality of the germ plasm was only relative had already been clearly intimated, however, in his more important work, Das Keimplasma, 1892, p. 526. Speaking of a butterfly, which has bright or dark wings, according to the climate, he goes on to say: "The modifying influence, here temperature, affects in each individual both the fundament of the wings—that is, a portion of the soma—and also the germ plasm contained in germ cells of the organism. In the wing-fundament the same determinants change as in the germ-cells—namely, those of the wing-scales. The first modification can not influence the germ cells, and so affects only the colour of the wings

the inheritance of acquired characters for at least the lowest orders, for the *mixing* presupposes some given differences.)

* This appears in the early essay Ueber die Vererbung, Jena, 1883, p. 49.

belonging to the one individual; but the other passes over to succeeding generations and determines the colour of their wings so far as they are not further modified by later temperature conditions" It is only by means of such variations in the germ structure, brought about by external influences, that Weismann can now find a possible explanation of the origin of new species.*

I now pause to gather up these positions. Weismann's theory is not sufficiently defined by the thesis: there is no inheritance of acquired characters; for, in the first place, he grants such inheritance in the case of unicellular structures without a nucleus, where his distinctions between *morphoplasm* and *ideoplasm, somatogen* and *blastogen* do not hold; and, secondly, while there is indeed for him no inheritance of acquired characters among individuals of the higher forms of life, there is the inheritance of the acquired characters of germ plasm. For conditions which influence an individual organism may take effect in the hereditary substance present in it and produce inheritable changes in that substance. Acquired variation in the individual may run parallel, under certain conditions, with acquired and inherited variation in the germ plasm, but is never the cause of it. They are simultaneous reactions from a third condition—namely, the external influence. So it appears that what is usually meant by the phrase inheritance of acquired characters—namely, the carrying over from one generation to another of acquired characters of the body—is actually excluded by Weismann's theory.†

* Keimplasma, especially pp. 512, 514, 516. See also G. J. Romanes's Examination of Weismannism.

† I can not resist citing Kant here as an advocate of the old preformation doctrine. In 1775 he said in his article on The

It is undoubtedly true that Weismann has seriously shaken the faith in inheritance of acquired characters which formerly played so important a rôle in philosophy, especially in the departments of ethics and sociology. He accomplished this quite as much by his searching criticism of the Lamarckian principle as by his own complicated theory of heredity. Even adherents of the Lamarckian system admit that its principles were rather too easily assumed. And, fortunately, one can speak of a neo-Darwinism as opposed to neo-Lamarckism * without being pledged to all the mysteries of Biophores and Determinants, Ides and Idants. Galton,† an author whose stirp theory is in many respects analogous, is very sceptical in regard to the inheritance of acquired characters, if he does not abso-

Various Races of Man: "Organic bodies naturally contain germs of special developments that pertain to special parts. Birds of the same species that live in different climates have the germs of an extra set of feathers, that are developed or not according to climatic conditions. . . . External things may be the occasion but can never be the cause of such developments, which are always hereditary and specific. Accident or mechanical-physical causes can as little effect any permanent modification in the form or attributes of the members as they can produce an organism itself. . . . Diseases are sometimes hereditary, but these do not belong to the organism but are rather ferments in bad humors which propagate themselves by infection. . . . Air, sunshine, and food may modify an animal's body during its growth, but these modifications are not to be confused with the generative force which carries on its operations independently of them, so that whatever was to be perpetuated was already being developed for the advantage and permanence of the creature." Kant here speaks of modifications within the species, and from a teleological standpoint. Nevertheless the similarity of ideas is astonishing.

* Lester F. Ward, Neo-Darwinism and Neo-Lamarckism, 1891.

† Francis Galton, A Theory of Heredity, Journal of the Anthropological Institute, vol. v, pp. 329 ff., and especially 344 f.

lutely deny it. Similar opinions are held by James, Virchow, Meynert, His, Ziehen, O. Flügel, Wallace, Ray-Lankester, Thiselton Dyer, Brooks, Baldwin, Van Bemmelen, Spengel, and many others.* A. Forel has also joined their ranks.† He says: "I, too, used to believe that instincts were hereditary habits, but I am now convinced that this is an error, and have adopted Weismann's view. It is really impossible to suppose that acquired habits, like piano playing and bicycle riding for instance (these are certainly acquired), could hand over their mechanism to the germ plasm of the offspring." ‡

The transition to the idea of instinct is easy at this point, for, even according to the latest formulation of Weismann's theory, it is quite impossible that the intelligent actions of ancestors should be transmitted to their descendants as instincts. Even if a modification of the hereditary substance should take place it would not originate in the intelligent act, but in the external conditions that impelled the individual to perform the act. Let us take Schneider's example of fear in the dark. Our ancestors frequently encountered in the dark the terrible cave bear. This repeated experience most probably produced in their brains an acquired sensory motor tract: "Dark—be wary!" Now, is it at all conceivable

---

* W. James, The Principles of Psychology, 1891, vol. ii, p. 678. Th Ziehen, Leitfaden der physiologischen Psychologie, 1893, p. 12. O Flügel Ueber das Seelenleben der Thiere, Zeitschrift fur exacte Philos., vol. xiii (1885), p. 143; Ueber den Instinct der Thiere mit besondere Rücksicht auf Romanes und Spencer (1890). ibid., p. 17. A. R. Wallace Darwinism. Baldwin, American Naturalist, June, July, 1896. For the other authors see Weismann's Das Keimplasma, p. 519.

† [As has Lloyd Morgan. See his Habit and Instinct.]

‡ A. Forel, Gehirn und Seele, 1894, p. 21.

that a variation parallel with this can have been effected in the reproductive substance through which predispositions arose that at once produced a similar tract in the brains of their descendants? This is incredible. It follows, then, since by supposition only the external environment and not the bodily changes work upon the germ plasm, that any explanation of instinct by means of the inheritance of acquired characters is quite impossible. There remains, then, to be considered in this example only the explanation by means of selection, the Darwinian position. This is simple enough, supposing it to be really a case of heredity (as I do not pretend to affirm categorically): it has always been the case that more of those individuals perished who were inclined to walk about carelessly in dark caves and woods.

Weismann himself has not neglected the question of instinct He said as long ago as 1883 that all instincts have their roots not in the acts of individuals, but rather in germ variation.* In the same lecture he also pointed out that many instinctive acts are performed only once in a lifetime—for instance, the flight of the queen bee—and would thus be inherited without practice.† And in a paper on the Allmacht der Naturzüchtung, 1893, he cites a highly interesting example which seems to exclude every explanation other than that of selection. It concerns, on the one hand, the origin of physical characters and of instincts, and on the other the decadence of the latter. In this case the inheritance of acquired characters is out of the question, as the subject is a sterile individual. The workers among ants are known to be sterile. Among

* Ueber der Vererbung. Jena, 1883, p 37.

† Ibid. See also the remark of Darwin cited above.

some species the female workers have the slaveholding instinct. This instinct must have arisen before the species had sterile workers (they have developed from females originally productive); for all the intermediate stages are known between those which hold no slaves at all and those which always do it. The *Formica sanguinea* do not yet show the slaveholding tendency as a fixed and demonstrable characteristic of their species, nor have they the extraordinary physical modification that marks the *Polyergus refescens*, settled slaveholders. Accordingly, we have here two developmental stages with clearly marked instincts. It is between these two stages that the variation to sterile workers must have taken place. The jaws must be changed from working tools to deadly weapons, as well as become adapted for carrying; they have become sword-sharp pincers, sharp and strong, suited alike for seizing and bringing home the young from other nests and for boring into the heads of enemies. At the same time the instinct for plundering is enormously strengthened. And here the hereditary effect of practice can not possibly be argued. The sterile workers could not possibly transmit anything, and their progenitors possessed neither such organs nor such instincts. On the other hand, the domestic instincts are weakened; workers of the *Polyergus* neither care for the larvæ nor collect food and building material. In fact, they seem to have lost entirely even the capacity to recognise and appropriate their own proper nourishment. "Forel, Lubbock, and Wasman are all convinced that the assertion made by Huber long ago is entirely correct. I have repeated his experiment, as well as Forel's, with the same result. These insects starve when confined, if none of their slaves are at hand to feed them. They do not recognise a drop of honey as something that will satisfy their hunger, and when Wasman placed a dead pupa actually in their jaws, they made no attempt to eat it, only licked it inquiringly, and left it. But as soon as one of their slaves—that is, a worker of the *Formica fusca*—was introduced, they approached and begged it for food. The slave hastened to the honey drop, filled her mouth, and brought the food to her ladyship." * What a splendid example this would be of the hereditary effect of disuse, says Weismann, if only these workers were not sterile! As the unreasoning conduct of these workers excludes the idea that their behaviour springs from the judgment of individuals, there remains for an explanation only selection, and selection of the mother at that. It must be noted, on the one hand, that those ant communities are more thriving whose productive females bring forth workers whose individual variations are in the direction of the decided modifications in physical qualities and in instinct mentioned above.† On the other hand, if the force of selection relaxes with reference to the weakening instinct, there results a community where the fruitful females produce workers whose instinct for collecting, rearing the young, and foraging is constantly diminishing (negative selection or panmixia).

Instances like this must increase the doubt about the inheritance of acquired characteristics. Let us now turn our attention to some other arguments for neo-Darwinism. To its advocates the fact is very significant that not a single example seriously threatening Weismann's theory has been brought forward by their op-

* Weismann, Der Allmacht der Naturzüchtung, Jena, 1893, p. 52.

† The same idea may be found in Darwin's Origin of Species, *in loc.*

ponents. Many of the cases cited with that in view are scientifically unreliable, and the rest can be explained quite well by the principle of selection. If acquired characters were hereditary, what an instinctive predisposition there would be for such acts as writing, for instance!—and Spencer would be able to ascribe Mozart's precocious musical talent to the practice of a few previous generations! It should further be borne in mind that the long-continued experiments of Weismann and others have never produced a positive instance of such transmission. Darwin himself was interested in the question. Romanes tells us that in 1874 he had a long conversation with him on the subject, and undertook a systematic series of experiments under his direction. He continued them for more than five years almost uninterruptedly, but they were all unsuccessful; so he, too, found it impossible to establish the truth of the inheritance of acquired characters.*

As regards instinct, there is, further, the *a priori* argument that it is inconceivable how acquired connections among the brain cells could so affect the inner structure of the reproductive substance as to produce inherited brain tracts in later generations. And, finally, there is this consideration mentioned by Ziegler as a suggestion of Meynert's: "It is well known that in the higher vertebrates acquired associations are located in the cortex of the hemispheres. As an acquired act becomes habitual, it may be assumed that the corresponding combination of nervous elements will become more dense and strong and the tract proportionally more fixed. This being the case, it follows that the tracts of acquired and habitual association, as well as those of

* Romanes, A Critical Examination of Weismannism, preface.

acquired movement, pass through the cerebrum. Instincts and reflexes, however, have their seat for the most part elsewhere. The tracts of very few of them are found in the cortex of the hemispheres. It is chiefly in the lower parts of the brain and spinal cord that the associations and co-ordinations corresponding to instincts and reflexes have their seat. When the comparative anatomist investigates the relative size of the hemispheres in vertebrates (especially in amphibians, reptiles, birds, and mammals), a very evident increase in size is observed which apparently goes hand in hand with the gradual gain in intelligence. In the course of long phylogenetic development, during which the hemispheres have gradually attained their greatest dimensions, they have constantly been the organ of reason and the seat of acquired association. If, then, habit could become instinct through heredity, it is probable that the cerebrum would in much greater degree than is the fact be the seat of instinct." *

But what part has psychology had in this war of opinions? It is impossible for her to give a satisfactory answer to this question. She must pick her way cautiously, and in the matter of instinct the adoption of the neo-Darwinistic theory is evidently the most prudent course, for to it belongs the now universally recognised principle of selection. Accordingly, when I speak of instinct it will be in accordance with this idea of innate hereditary variations, passing by the Lamarckian theory as either obsolete or a point of view yet to be substantiated. In what follows I adhere in essentials to the definition that Ziegler, a follower of Weismann, has given in

* H. E. Ziegler, Ueber den Begriff des Instincts, Verh. d. deutsche zool. Gesellschaft, 1891, p 134. See also Baldwin, Mental Development in the Child and the Race, chap. vii, § 4.

the address already cited, with the exception of one point, which I shall indicate at once. Ziegler has set forth with great skill, clearness, and technical scholarship a point of view which is now more and more attracting the attention of modern zoölogists; and the leading features of his exposition coincide with the views of many modern psychologists. In all instinct there is a close connection between a particular stimulus and a particular act, a connection that is of utility under ordinary conditions. Is this useful adjustment attributable to conscious will? No. On the contrary, the idea of consciousness must be rigidly excluded from any definition of instinct that is to be of practical utility. (Who can tell whether a dog, a lizard, a fish, a beetle, a snake, or an earthworm performs an action consciously or unconsciously? It is always hazardous in scientific investigation to allow a hypothesis which can not be tested empirically.)

It follows, that such fixed and useful connections between stimulus and action are to be treated as reflexes. Instincts are, as Herbert Spencer has rightly said, complex reflex acts. But the connection between reflex action and instinct is explicable only by means of selection, and selection in the Weismannian sense, which excludes the inheritance of acquired characters. "In the progress of phylogenetic development natural selection lays the foundation of instincts, and accordingly they are useful. Instincts are adapted to conditions, and serve generally for the preservation of the individual, always for that of the species." There must be, physiologically speaking, certain connecting paths among the ganglion cells that—existing as hereditary predispositions—contain "hereditary tracts."

* See James, The Principles of Psychology, ii, p. 391.

The complexity of instinct that is often so baffling, and its wonderful adaptiveness are, after all, not more difficult to explain than the other things about an organism. For example: "The marvellous instinct that leads the wood bee (*Xylocopa violacea Fabr*) to build its intricate nest in the trunks of trees is not more inexplicable than the faceted eyes of these very insects. . . . The principles involved in the morphological structure of the organ also account for the instincts; and there are also to be taken into account homology, analogy, and parallel development, individual variation, natural selection, and the resulting adaptation, cross-breeding, and atavism; here also there are cases of rudimentary and hindered development, natural or artificial deformity." No part can be had, in the genesis of instinct, by association resting on a foundation of previous experience, what we mean by understanding, intelligence in its widest sense; nor by acquired tracts, for these are not hereditary.

After giving this elaboration of Ziegler's theory in his own terms, I make these essential points:

1. The assumption that intelligent acts are the ground of the origin of instinct is unwarrantable. Even if the Lamarckian theory is not absolutely tabled, it is much wiser, so long as the question remains open, to be content with the leading Darwinian principle, since its grounds are more assured.

2. In the explanation of instinct (and of play) we need consider only natural selection, for we do not know any other principle of development. The simple reflex action must develop in the process of time into the complex reflex actions that we call instinctive. In this way we try to explain their adaptability as well as we explain organic adaptability in general. Whether it can be satisfactorily done is another question. I am not one of the

number who believe in the "all-sufficiency of natural selection." Leaving out of the question the fact that our knowledge of phylogenesis rests finally on the mysterious ocean of metaphysical problems, of which I have spoken, it is by no means settled, even in the sphere of empirical science, that selection of ordinary individual variation suffices to bring about, even gradually and by minute degrees, a change from one species to another. There are those who deny this, to whom the Darwinian system is comparatively insignificant. As in surging water the particles of each wave move both backward and forward, so that the surface motion forward is really only apparent, so the selection of hereditary qualities can not extend beyond a certain definite point, and for the transformation to new species other and essentially different variations are necessary, in their opinion, in the structure of the germ substance itself.*

Nevertheless, we know no principle except that of selection, and we must go as far as that will take us. Absolute knowledge of such phenomena is practically unattainable.†

* Thus Galton and Bateson. F. Galton, Discontinuity in Evolution, Mind, 1894.

† Since this was written a new theory has been proposed which is evidently well adapted to supplement the selection principle. Baldwin has discovered a way whereby natural selection is furthered by individual accommodations or functional adaptations, and directed by them without the assumption of any direct inheritance of acquired characters; as he says "the appearance of such inheritance will be fully explained" (Mental Development in the Child and the Race, German translation, p. 188, fourth English edition, chap. vii, § 4). [*Cf.* The Psychological Review, vol. iv, p. 394 f., July, 1897. This influence is called by him "organic selection".] Independently of Baldwin, Osborn and Lloyd Morgan have reached a similar position. [It is also now accepted by Mr. Alfred Russel Wallace and Prof. E. B. Poulton, of Oxford. The latter says (Science, New

3. Since instincts, according to Spencer's view, already explained, are only complicated reflex acts, the question may be excluded whether animals acting instinctively are conscious in play of what they do. It is evident, of course, that many instinctive actions are accompanied by consciousness, but seeing that even the instincts thus consciously practised are probably derived from unconsciously perfected reflexes, it is impossible to draw the line.*

Ziegler is more cautious than Romanes and Schneider, who attempt to find a definite boundary line between

---

York, October 15, 1897, p. 585): "These authorities justly claim that the power of the individual to play a certain part in the struggle for life may constantly give a definite trend and direction to evolution, and that although the results of purely individual response to external forces are not hereditary, yet indirectly they may result in the permanent addition of corresponding powers to the species. . . . The principles involved seem to constitute a substantial gain in the attempt to understand the motive forces by which the great process of organic evolution has been brought about."] The importance of this theory seems to me to depend mainly upon whether the fostering of "congenital variations" in this way is of sufficient "selective value," even though we grant the supposition made, that the animals are kept alive by their individual accommodations. Baldwin has considered this point (Science, New York, March 20 and April 10, 1896), but perhaps without giving it sufficient prominence. It should also be borne in mind that individual accommodations become through practice instinct-like ("semi-automatic"), so that the necessity for the perfection of the congenital function is somewhat diminished. [*Cf.* the Appendix.]

* W. Wundt (Grundzüge der physiologischen Psychologie, ii, 582) and E. Alix (L'esprit de nos bêtes, Paris, 1890, p. 580) are of the opinion that the reflex itself is conscious movement become mechanical. Both of them connect this idea with the inheritance of acquired characteristics. Apart from that, it is not contradictory of what has been said, for frequently-repeated reflexes are often carried on unconsciously, even when they were accompanied by consciousness when first appearing in the individual.

instinct and reflex action according as consciousness is present or not.* In the opposite direction, he is more cautious than Ziehen,† who accepts the hypothesis of the absolute unconsciousness of instinctive acts. Ziegler is probably influenced here, as on other points, by Herbert Spencer, who thus guardedly expresses himself: "Instinct in its higher forms is probably accompanied by a rudimentary consciousness."‡ So far I agree with Ziegler, but his avoidance of any definite expression of opinion as to whether consciousness is or is not present is significant in another connection, and here, as I think, he is not entirely in the right. Every instinctive act is a means for preserving the species. This fact gives the question of consciousness a double significance, as Hartmann's definition, for example, clearly shows: "Instinct is the conscious willing of means to an unconsciously willed end." #

As concerns the means, that is, the act itself, it is safer, as has been remarked, to avoid the terms "conscious" and "unconscious" altogether. But it seems permissible to say, at least with reference to the end of a particular action, "by instinct we understand the impulse to an action whose end the individual is unconscious of, but which nevertheless furthers the attainment of that end." ‖ That is to say, the consciousness of an end as such is entirely separable from the instinctive act. Ziegler does not leave room for any psychic factor, not even a negative one, in his definition.

* Romanes, Animal Intelligence, 1892, p. 11. Especially in Schneider's book, Der thierische Wille.

† Th. Ziehen, Leitfaden der physiologischen Psychologie, p. 12.

‡ H. Spencer, Principles of Psychology, p. 195.

Hartmann, Philosophie der Unbewussten, i, p. 76.

‖ Schneider, Der thierische Wille, p. 61.

"Who can know whether the bird when she builds her nest already has the knowledge that her young will find a warm bed in it? And even as applied to man this criterion is misleading. For example, when a mother suckles her child the action is evidently instinctive, though the mother perhaps cherishes the hope that the child may become the support of her old age and the representative of his family, thus knowing perfectly well not only the immediate end of her action, but also its utmost consequences." On this account Ziegler prefers not to speak of the presence or absence of consciousness of end or object. But it seems to me that the subject has quite a different aspect if we first try to make clear just what is meant by lack of consciousness of end or object. There are two widely different ways of interpreting the expression. First, there is the relativity of the end to be considered, as Schneider has justly remarked in his later work, Der menschliche Wille.* When a beast of prey scents his victim, and creeps toward it with the movement peculiar to his kind, this creeping is a means to the end of approaching near enough for a spring. The spring is a means to the end of seizing the animal and slaying it. Rending the prey is a means to the end of eating it, and this in turn serves the end of nutrition, and so on. Only the last and highest end is, as far as we know, not a relative one—namely, the preservation of the species. But under present conditions only reflecting man can be conscious of this end, and even he is scarcely conscious of it in actual everyday life. There is usually only a relative consciousness of end even in our actions which are not instinctive. When a man buys a new suit

* Darwin also, in The Origin of Species, p. 328.

of clothes he does not reflect that he is thus furthering the preservation of his kind (Schneider). As for the instinctive acts of children, savages, and animals, it may safely be affirmed that in them such adaptation of means to the end as selection requires for the preservation of the species is entirely unconscious. At the same time there may very well be consciousness of relative ends. The fox out hunting, for instance, may have a memory of gastronomic enjoyment in his mind as his end idea. I consider this first conception of activity unconscious of its end inadequate, however, because, as has been said, actions not instinctive are also often unaccompanied by a consciousness of their highest or final end.

Nevertheless, the position to be mentioned as second, combated by Ziegler, seems to me to be nearer the truth, namely, the position that an action is only instinctive when it does not include a consciousness of end, either relative or absolute, as its motive. Let us again take the fox, scenting his prey. If in creeping toward it he has a conscious end, this can only be grounded in individually acquired associations of smell with the agreeable taste of the victim, and in the recollection that it has been known to escape in consequence of careless movements on the part of the pursuer. We can not speak of instinct within the limits of such acquired association, so far as it operates as a motive. So far, on the other hand, as the mere external stimulus to the olfactory nerves of the fox excites to functional activity the hereditary tracts in the animal's brain, so far his act is just as instinctive as the spitting of a kitten at the hand which has stroked a dog, or the bird that builds a nest. Even if the bird does have the consciousness that its young will find a warm bed there, its action may still be purely instinctive, so long as that consciousness remains a mere memory, without motive power. As soon, however, as the idea affects the will, we have no longer a purely instinctive action to deal with, but one that is partly instinctive and partly voluntary. Inasmuch, also, as conscious action often tends to become instinctive, I may take account of that fact, and accordingly formulate this approximate definition: The actions of men and animals are instinctive when originated by means of hereditary brain tracts (presumably of selective origin) and without an idea serving as their motive.

The fact that the same act may be partly instinctive and partly voluntary is of importance in many connections, not least in that of play, in which the higher the stage the more the individual accommodations are involved. Formerly extreme theorists entertained the view that only animals have instinct, only man has reason. Cuvier believed that the relation of instinct and intelligence was that of inverse ratio; Flourens, the same. Darwin opposes Cuvier's idea, but thinks that "man perhaps has somewhat less than the animals standing next him," and that the instincts of higher animals are less numerous and simpler than those of the lower orders.* James, on the contrary, reverses the proportion, and says that man is probably the animal with most instincts.† This is perfectly true if it is borne in mind that some actions are partly voluntary and partly instinctive. Take, for instance, lovers of the

* Descent of Man, ed. in one vol., p. 75.

† The Principles of Psychology, vol. ii, p. 389. Pouchet, too, in the Revue des deux mondes, February, 1870, and Alix (L'esprit de nos bêtes, 1890) express the view that the most intelligent animals have an especially large number of instincts. [See also Lloyd Morgan, Habit and Instinct, for a general discussion and (p. 328) for a criticism of James. The same author collects and criticises Some Definitions of Instinct, in Natural Science, May, 1895.]

chase, who are perfectly conscious of the object of their actions and yet are in great part impelled by instinctive impulses. If such half-instinctive phenomena are included in the category, then man has as many instincts as any animal, if not more. By this elucidation we reach the truth that lies concealed in the theory mentioned above—namely, that the lower the animal stands the purer are its instincts; the higher its place the more will the hereditary tracts be weakened, altered, or supplanted by acquired tracts. "The more various and ready the inherited mechanical impulses of a class of animals," say the Müllers, "the less do we find of independent mental capacity." * And Flourens remarks, "Intelligence does not enter into instinct, but it influences it, protects it, and alters circumstances to suit it, and this agreement between instinct and intelligence is well worth attention." †

I am now firmly convinced that this relation is itself eminently useful, and that it is due to negative as well as positive selection. Hartmann has already pointed out that Nature substitutes instinct where the means are not at hand for conscious action or acquisition.‡ The higher and more complicated the scale of activity which the struggle for existence requires of a species the more will selection favour development of the brain and of the mental capacities. The more these increase by means of positive selection the less will its aid be needed in the sphere of instinct. The result will be that fewer individuals will have completely developed hereditary tracts

* A. and K. Müller, Wohnungen, Leben und Eigenthümlichkeiten in der höheren Thierwelt, p. 217.

† P. Flourens, Psychologie comparée, second edition, 1864, p. 10. See also J. Sully, The Human Mind, 1892, i, p. 137.

‡ Hartmann, Philos. d. Unbewussten, i, p. 185.

for future transmission. In short, where positive selection furthers the growth of intelligence, for instinct there will be a certain degree of negative selection or panmixia. (This, of course, applies only to instincts for which conscious actions can be advantageously substituted.*) Indeed, it might even be said that the degeneration of instinct is due to positive selection. We have no intimation at what stage of evolution the animal world first achieves activity that depends on its own intelligence or the capacity for individually acquired association; but we may assume that at some point in the progress of evolution the creature attained sufficient intelligence to accomplish many things by means of it better than by instinct. From this moment on, extensive inheritance of brain mechanism would have been positively prejudicial to the further development of intelligence, and a positive selection may be assumed that would directly favour less finished instincts in order to produce in the nervous system a partiality for the now more useful acquired functions.†

* [Romanes thinks the existence of both sorts of function shows the inheritance of acquired characters in the case of instincts (Heredity and Utility, p. 74 f); but Baldwin shows that in such cases the instinctive performance has an additional utility, thus supporting the position of the text (Science, New York, April 10, 1896).]

† Wundt also points to this idea in the section on "Affects and Impulses" in his Physiolog. Psychologie. Vol. ii, p. 512, of the fourth edition runs thus: "The many-sidedness of a creature offers a wide field for individual development, and at the same time the *determination by heredity* is constantly diminished." *Cf.* the very precise statement of this by Baldwin in Mental Development in the Child, etc., chap. xvi, § 1 (German edition). [See Psychological Review, iv, July, 1897, p. 399, and his preface to this work.] The contrary supposition that imperfect instincts betoken early stages in evolution is surely incorrect in many cases.

Be that as it may, we may explain by such degeneration of instincts the countless cases which have caused such men as Wallace to doubt whether there is any instinct at all. In his chapter on The Philosophy of Birds' Nests, Wallace has collected observations intended to prove that birds do not come into possession of their songs by inheritance, but learn them individually. Barrington caged young linnets with singing larks, whose song they learned so well that, even when placed with other linnets, they did not change them. A bullfinch sang like a wren and without any of the characteristics of its own kind, and similar results were obtained from the wheat-ear, fallow-finch, nightingale, and woodpecker. "These facts," says Wallace, "and many others which might be quoted, render it certain that the peculiar notes of birds are acquired by imitation, just as a child learns English or French, not by instinct, but by hearing the language spoken by its parents." * This sounds very convincing, but it is first to be considered that the use of the voice is instinctive, and then that imitation itself is instinctive, of which more is to be said below; and, finally, that the experiment failed with young birds taken from the nest when only a few days old, for they could never be influenced again in the same way by later experiences. The song of birds is no doubt a mixed phenomenon in which instinct and experience blend.†

Such advancement of the evolution of intelligence as we have been considering is favoured also by play, as I

* Wallace, Contributions to the Theory of Natural Selection, *in loc.*

† [This conclusion is strongly supported by the researches of Lloyd Morgan. See his Habit and Instinct.] There is evidence, too, that complicated songs are produced without teaching. Simple calls like those of the cuckoo and quail are purely instinctive.

believe. I trace the connection as follows: A succession of important life tasks is appointed for the adult animal of the higher orders, as for primitive man, some of the principal being as follows:

1. Absolute control of its own body. Grounded on this fundamental necessity are the special tasks, namely:

2. Complete control over the means of locomotion for change of place, characteristic of the species, as walking, running, leaping, swimming, flying.

3. Great agility in the pursuit of prey, as lying in wait, chasing, seizing, shaking. Equal fitness for escaping from powerful enemies, as fleeing, dodging in rapid flight, hiding, etc.

4. Special ability for fighting, especially in the struggle with others of the same kind during courtship, etc.

After the foregoing discussion there can be no doubt that instinct plays a part in all this adaptation for the struggle for life and preservation of the species, so necessary in man and other animals. Further—and here I again come into touch with the end of the last chapter—it would be entirely in harmony with other phenomena of heredity if we found that these instincts appear at that period of life when they are first seriously needed. Just as many physical peculiarities which are of use in the struggle for the female only develop when the animal needs them; just as many instincts that belong to reproduction first appear at maturity; so the instinct of hostility might first spring up in the same manner only when there is real need for it; and so it might be supposed with other instincts in connection with the related activities. The instinct for flight would only be awakened by real danger, and that of hunting only when the animal's parents no longer nourished it, and so on. What would be the result if this were actually the case—if, in other words, there were no such thing as play? It would be necessary for the special instincts to be elaborated to their last and finest details. For if they were only imperfectly prepared, and therefore insufficient for the real end, the animal might as well enter on his struggle for life totally unprepared. The tiger, for instance, no longer fed by his parents, and without practice in springing and seizing his prey, would inevitably perish, though he might have an undefined hereditary impulse to creep upon it noiselessly, strike it down by a tremendous leap, and subdue it with tooth and nail, for the pursued creature would certainly escape on account of his unskilfulness.*

Without play practice it would be absolutely indispensable that instinct should be very completely developed, in order that the acts described might be accurately performed by inherited mechanism, as is also the case with such instinctive acts as are exhibited but once in a lifetime. Even assuming this possibility, what becomes of the evolution of higher intelligence? Animals would certainly make no progress intellectually if they were thus blindly left in the swaddling-clothes of inherited impulse; but, fortunately, they are not so dealt with. In the very moment when advancing evolution has gone so far that intellect alone can accomplish more than instinct, hereditary mechanism tends to lose its perfection, and the "chiselling out of brain predispositions" † by means of individual experience becomes more and more prominent. And it is by the play of children and animals alone that this carving out can be properly and perfectly accomplished. So natural selection

* In such a case, of course, the parents would never have brought him living prey to play with.

† E. v. Hartmann, Philosophie des Unbewussten, iii, p. 244.

through the play of the young furthers the fulfilment of Goethe's profound saying: "What thou hast inherited from the fathers, labour for, in order to possess it."

At this point the full biological significance of play first becomes apparent. It is a very widespread opinion that youth, which belongs, strictly speaking, only to the higher orders, is for the purpose of giving the animal time to adjust itself to the complicated tasks of its life to which its instincts are not adequate.* The higher the attainment required, the longer the time of preparation. This being the case, the investigation of play assumes great importance. Hitherto we have been in the habit of referring to the period of youth as a matter of fact only important at all because some instincts of biological significance appear then. Now we see that youth probably exists for the sake of play. Animals can not be said to play because they are young and frolicsome, *but rather they have a period of youth in order to play;* for only by so doing can they supplement the insufficient hereditary endowment with individual experience, in view of the coming tasks of life. Of course this does not exclude other grounds, physiological ones, for instance, for the phenomenon of youth; but so far as concerns the fitting of the animal for his life duties, play is the most important one.

I may now briefly recapitulate. Our leading question seems to be as to the play of the young. That once adequately explained, the play of adults would present no special difficulties. The play of young animals has its origin in the fact that certain very important instincts appear at a time when the animal does not seri-

* See J. Mark Baldwin, Mental Development in the Child and the Race, 1895, p. 28 f.

ously need them. This premature appearance can not be accounted for by inherited skill, because the inheritance of acquired characters is extremely doubtful. Even if such inheritance did have a part in it, the explanation by means of selection would still be most probable, since the utility of play is incalculable. This utility consists in the practice and exercise it affords for some of the more important duties of life, inasmuch as selection tends to weaken the blind force of instinct, and aids more and more the development of independent intelligence as a substitute for it. At the moment when intelligence is sufficiently evolved to be more useful in the struggle for life than the most perfect instinct, then will selection favour those individuals in whom the instincts in question appear earlier and in less elaborated forms—in forms that do not require serious motive, and are merely for purposes of practice and exercise—that is to say, it will favour those animals which play. Finally, in estimating the biological significance of play at its true worth, the thought was suggested that perhaps the very existence of youth is largely for the sake of play.

The animals do not play because they are young, but they have their youth because they must play.*

But I must call attention to another important phenomenon that also has a direct relation to play, namely, the imitative impulse. It was remarked in the previous chapter that while imitation is not an essential feature of play, it is very often present. This is a suitable place to notice this important subject, which will constantly recur in the progress of our inquiry. First, it

* [The author here adds several pages in which he suggests that the conscious accompaniments of play—fully described in later chapters—are also due to natural selection; and points out the "play comradeship" of young animals, saying, "Daher ist die sociale bedeutung der Spiele ausserordentlich gross."]

is very probable that imitation is itself instinctive. True, it is possible to conceive of the imitative impulse as of individual origin. Wundt teaches that every idea of movement presses to fulfil itself. (Many psychologists seek to reduce even the will to such ideas.) The notion of the movements seen in others is, of course, included, and this is the imitative impulse.* But an origin so entirely individual and even accidental can hardly be attributed to an impulse of such enormous power. Wundt refers to the impulses, too, as hereditary phenomena, and, if I understand him aright, does not exclude imitation.† Schneider thus expresses himself on the subject: "Wundt is quite right in regarding apperception of a movement idea, and the feelings connected with it as a direct impulse to make the movement. And the word idea is not used in a narrow sense, for even the perception of a movement awakens this impulse, and is the cause of many imitative movements." Schneider is, however, of the opinion that the development of this "intimate causal connection" rests in both cases on heredity (according to him, indeed, on the inheritance of acquired characters), and advances as an explanatory proof of this the fact that the imitative impulse is restricted to cases that are useful to the individual. "When a young lion sees a fish swimming or a bird flying he hardly feels a desire to swim or fly, while the old lion's movements when he observes them arouse the imitative impulse in him, because he is disposed to the movements by heredity. This is a proof that

* Wundt, Phys. Psych., fourth edition, p 567 ff. The same thought is brought out by James Mill in his Anal. of the Phenom. of the Human Mind, 1829, vol ii, chap. xxiv.

† See Vorlesungen über Menschen- und Thierseele, second edition, p. 433. [Eng. trans., pp. 388 ff.]

apperception should not be regarded as *only* a movement idea, for if that were all it is, animals, at least, would seek to imitate every motion they see, and we might expect to see a child at once begin a swaying motion on beholding a pendulum, instead of reaching for the ball of it, as it does." *

Spencer also, James, and Stricker, regard the imitative impulse as an inherited instinct, and I think it is safe to trust these psychologists.†

The imitative impulse is thus found to be an instinct directly useful in the serious work of life among most, and presumably among all, of the higher gregarious animals. Its simplest manifestation is the taking to flight of a whole herd as soon as one member shows fear.‡ A more particular case is seen among certain domestic animals that blindly follow a leader—a fact well known to the crafty Panurge in Rabelais's grotesque romance, when during the voyage he wants to play a joke on the owner of a flock of sheep.# This phenomenon, which may be explained by the principle of division of labour (for in this way one animal can watch for the whole

* Der menschliche Wille. p. 311.

† Sully's ground for combating this view (The Human Mind, 1892, vol. ii, p. 218), namely, that imitation first appears in a child's fourth month, is, of course, no argument against the heredity of the impulse. [Baldwin has published a special criticism of the arguments against the instinct-view of imitation, *loc. cit.*; and Bain, who formerly did not accept the instinct-view, has recently adopted it. *Cf.* Baldwin, Mind, January, 1894, p. 52; and Bain, Senses and Intellect, fourth edition, p 441.]

‡ According to Wallace, the white back of many animals is a signal of danger for inciting their comrades' imitation in flight. Darwinism, pp. 217 ff.

He persuaded the owner to sell him a sheep, chose the bell-wether, and threw him overboard, whereupon the whole flock jumped after him and were drowned.

flock), is also useful in advancing intellectual development, since selection favours young animals skilled in its use. So we have here an hereditary instinct that is even more especially adapted than that of play to render many other instincts unnecessary, and thus open the way for the development of intelligence along hereditary lines that can be turned to account for the attainment of qualities not inherited. Young animals, even some not gregarious, have an irresistible impulse to imitate any action of their parents, toward which their instinctive impulse is very weak, and they learn in this way what would never be developed in them individually without this imitative impulse. The examples cited from Wallace can be explained in this way. They do not argue against instinct, but rather show that many instincts are becoming rudimentary in the higher animals because they are being supplanted by another instinct—imitative impulse. And this substitution is of direct utility, for it furthers the development of intelligence. This reminds us of the teaching of Plato, that the ability to learn presupposes " reminiscence " from a previous existence. By means of imitation animals learn perfectly those things for which they have imperfect hereditary predispositions

We then reach the following conclusion in our play inquiry—namely, that all youthful play is founded on instinct. These instincts are not so perfectly developed, not so stamped in all their details on the brain, as they would have to be if their first expressions were to be in serious acts. Therefore they appear in youth, and must be perfected during that period by constant practice. At the same time, where physical movements are concerned, the muscular system will also be developed by this exercise suitably for subsequent serious work—a result which would not be attained adequately without play. In this way we can explain those plays referred to in this and the previous chapter, which can not be designated as imitative play, such as the gambolling of young creatures, their play with the organs of motion and speech, mock fighting, etc. Besides these plays, which are founded on strongly developed instincts, and can therefore be practised without a model, there are many others worthy of consideration: those in which at least two instincts are involved—one an impulse only rudimentarily present, though easily aroused, and the other the accompanying imitative instinct To this class belong the instances already cited of young birds learning to sing, probably, too, the barking of puppies, and the imitative play of little girls whose motherly tending of their dolls could hardly reach the perfection in which we see it without imitation. It would be certainly hard to explain the choice of models by the different sexes without hereditary predisposition—why the boy's tin soldiers are his favourite toys, while the little girl is always the mother and housekeeper. Finally, it must be admitted that there are cases where the imitative impulse exceeds the limits of instinct and apparently works alone, as when apes imitate the actions of men, when parrots learn to speak intelligently, and when children play horse cars, railroad, hunter, teacher, and the like. But even here a latent desire to experiment contributes, and it is evident how necessary such play is to the development of mind and body.

We now have all the principles necessary for a psychology of play; only in outline, however. All refinements and expansions which may subsequently be brought to light, and which I may call idealizations of the bald play instinct, must be treated later. The following remarks will conclude this chapter:

Play is found among adult animals. A creature that once knows the pleasure of play will derive satisfaction from it even when youth is gone. And preservation of the species is advanced by exercise of the mind and body even in later years. I have a dog twelve years old that still shows a disposition to play now and then. We often see grown-up animals playfully roll over and over without any object, and many birds appear to sing from mere sportiveness without relation to courtship. Proof of this is difficult to substantiate, however. We do know that adult cats and dogs play, but in regard to other animals we can only speak of probabilities. If the playful character of some of the examples which I cite in the following chapters is not established beyond a doubt, I am consoled by a statement of Darwin's, made with great emphasis in The Descent of Man: * "Nothing is more common than for animals to take pleasure in practising whatever instinct they follow at other times for some real good. How often do we see birds which fly easily, gliding and sailing through the air obviously for pleasure? The cat plays with the captured mouse and the cormorant with the captured fish. The weaver-bird when confined in a cage amuses itself by neatly weaving blades of grass between the wires of the cage. Birds which habitually fight during the breeding season are generally ready to fight at all times, and the males of the capercailzie sometimes hold their *Balzen* at the usual place of assemblage during the autumn. Hence it is not at all surprising that male birds should continue singing for their own amusement after the season for courtship is over."

---

* Descent of Man, vol. ii, p. 60.

CHAPTER III.

THE PLAY OF ANIMALS.

THE following treatise forms, so far as I know, the first attempt at a systematic treatment of the play of animals, and, in view of the unavoidable difficulties inherent in the task, I wish to bespeak the reader's indulgence at the outset. Modern works on the mental life of animals, such as the writings of Carus, Schneider, Wundt, Buchner, Espinas, Romanes, Lloyd Morgan, Flourens, Alix, and Foveau de Courmelles, contain only meagre and general accounts of even the most important plays.*

Thus Romanes, in his laborious work, Animal Intelligence, which in the edition of 1892 numbers five hundred pages, makes, aside from the play of ants and dolphins, only a few incidental observations on the play of birds, dogs, and monkeys.†

The great significance of play in physical and mental development seems not to have attracted the attention of psychologists as it deserves to do. Therefore I hope that this book, in spite of its imperfections, may contribute to the result that in the future every ani-

---

* Among older works, Scheitlin's Thierseelenkunde is famous.

† The observations made by this author's sister on a young ape, included in the book, are much more valuable.

mal psychology shall contain a chapter devoted to play.*

On account of this defect in the specific works on animal psychology I have been obliged to seek for most of my material from other sources, and especially from such books as contain descriptions of the habits of animals, though without aiming to meet the requirements of psychology. Most of the observations described are from Naumann, Bechstein, Rengger, Lenz, Ch. L. and A. E. Brehm, K. and E. Müller, Tschudi, Russ, Diezel, Marshall, Darwin, Miss Romanes, Wallace, and Hudson. A. E. Brehm's *Thierleben* is the richest of these. It is marred by the attempt to humanize the actions of animals, but this defect is not injurious to his descriptions of plays. The examples without references in this and the next chapter are from it. I have also made use of such periodicals as *Gartenlaube* and *Der Zoologischen Garten*. I have examined many books of travel, but usually with discouraging results. If they refer to the play of animals at all, the most they say is that is was "amusing," or "astonishing," or "droll," or "exceedingly funny," without any account of how or why. Such a description as that of the young gorilla and some other animals in the "Loango Expedition" forms a notable exception. As far as personal observation goes, I am familiar with the habits of dogs, as I have always from my youth had various breeds of them about me; and also I have collected enough material in my frequent visits to the zoological gardens to furnish cases of some kinds of play from my own observation. A complete review of all animal plays is not

* Alix alone gives it a single paragraph, and that is, of course, totally inadequate.

possible here; indeed, I have confined myself in essentials to phenomena from the life of the higher orders, because the play of the lower ones seems to me to be too little known. I have multiplied examples in those departments where errors of judgment are most liable to occur, and can only be set right by such fulness of detail. I am afraid that this result has not always been accomplished, however, and in the case of the so-called love-plays the material was so copious as to compel me to suppress much that was interesting.

There is no difficulty in classifying our subject if the conception developed in the preceding chapters is accepted. I hope that no essential group has been left out of the following table:

1. Experimentation.
2. Movement plays.
3. Hunting plays:
    a. With real living prey.
    b. With living mock prey.
    c. With lifeless mock prey.
4. Fighting plays:
    a. Teasing
    b. Tussling among young animals.
    c. Playful fighting among grown animals.
5. Love plays:
    a. Among young animals.
    b. Rhythmical movements.
    c. The display of beautiful and unusual colours and forms.
    d. The production of calls and notes.
    e. The coquetry of the female.
6. Constructive arts.
7. Nursing plays.

8. Imitative plays.
9. Curiosity.

This arrangement will be followed in order throughout, except that I have treated love plays, which deserve more than superficial elaboration, in a separate chapter, after all the others.

1. *Experimentation.*

In opening this subject we are at once confronted by a group of phenomena, familar enough in children, but hardly noticed heretofore in the psychology of animals. The term experimentation is here used to denote such movements of young animals as enable them first to win the mastery over their own organs, and then over external objects. It includes stretching and straining the limbs; tasting, seizing, and clawing; gnawing and scratching; exercising the voice and making other sounds; rending, pulling, tearing, tugging, kicking, lifting, and dropping objects, etc. Such experimental movements are of fundamental importance for all the life tasks of animals, for on them depends the proper control of the body, muscular co-ordination, etc.; and, psychically, they promote the development of the perceptive faculties, such as space perception, attention, will power, memory, etc. They form the common foundation on which the specialized plays are built up. Though the term hardly seems quite applicable to all the examples included under this heading, I use it in default of a better. It seems to have originated, so far as I can trace its use, with Jean Paul, who speaks in his Levana of "the child's experimental physics, optics, and mechanics" He says, "Children take the greatest pleasure in turning things around, in lifting them, sticking keys in locks or anything of the sort, even in opening and shutting doors." * Later, B. Sigismund made use of the expression in his serviceable little book,† and Preyer and Sikorski have established its use in modern psychology.

Since the babyhood of animals is so much shorter than that of the human infant, it offers much less material for psychological investigation, and, besides, there is no Preyer for the animals.‡ Still, we are not entirely without material.

"With the stretching of his limbs," say the Müllers, "the young dog begins the first stage of his baby play."# Puppies also begin very early to gnaw any wooden object, as well as their own extremities, with their little teeth, sharp as needles. Even the play with their tails is at first purely experimental. Afterward the chase instinct comes in, when the end seems to vanish so mysteriously as they whirl. A dog that I once owned was so small and weak that he always tumbled over in attempting to bark. It was most ludicrous to witness this ignominious ending to his hostile demonstrations. A kitten, too, will play with its tail, and exercise the claw-armed paws in seizing and holding.‖ Scheitlin observed a young panther playing with its own tail,△ and Brehm relates how pumas at the age of

---

* Jean Paul. Levana 2d ed., vol. i. p. 164.

† B. Sigismund, Kind und Welt, 1856, p. 73.

‡ The thoroughgoing papers of Wesley Mills on the psychic development of young animals were not known to the author when this passage was written. They were published in 1894—1896. [Prof. Lloyd Morgan's observations on young birds should also be referred to; *cf.* his Habit and Instinct, 1869.]

\# A. and K. Müller, Charactere aus der Thierwelt—1. Der junge Hund, Gartenlaube, 1867, p. 455.

‖ As Wesley Mills observed on the twenty-ninth day. *Loc. cit.*, part ii.

△ Thierseelenkunde, ii, p. 155.

from five to six weeks play with their mother's tail, as do all the cat tribe. He also tells of a young fish-otter that snapped at its tail and fore paws. This, however, appears to belong rather to the chase phenomena, as it is not purely experimental. But there are no clearly defined boundaries between general experimentation and specialized plays. The cat observed by Wesley Mills touched the poker (on its fifty-ninth day), which was hot. It hissed, but soon after it touched it again "in its usual persistent way." It was fond of knocking down spools from the table, and especially delighted in taking pins out of the cushion.

A young polar bear that I knew often lay on its back and bit its paws, or tried to tear a piece of paper, and it has frequently been noticed that young bears make a humming kind of sound, ending with a smack, when they suck their paws.* Falkenstein relates of his gorilla, about a year and a quarter old: "He delighted in the bath, and after a while tried to help himself when I did not appear at his side at the right moment with sponge and soap. That the water all ran out of the tub in a few moments did not affect his enthusiasm. He paddled on all fours in the wet, like the little darkies during a shower." †

Little nestlings make fluttering efforts before they can fly, and young sparrows chirp so lustily in the nest as to suggest genuine voice practice. "Immediately on being hatched," says Hermann Muller, "the young birds begin to lift up their voices. Of canaries, goldfinches, siskins, and bullfinches hatched in confinement, canaries peeped earliest and loudest, bullfinches latest

* L. Brehm, Bilder aus dem Thiergarten in Hamburg—2, Unsere Bären, Gartenlaube, p. 12.

† Falkenstein, M-pungu, Gartenlaube, 1876, p. 556.

and weakest, suggesting that the birds' later capacity for singing might be gauged by their first twittering These loud, piercing notes are by no means signs of hunger, but, on the contrary, indicate the greatest contentment, for they cease at once when the mother leaves them and cool air fills the nest."

I must insert here a remark that belongs to the idealization of play. We may safely assume that the satisfaction of instinctive impulse is not the only pleasure in experimentation. Even in the animal intelligence it denotes a finer psychical state. Preyer calls the satisfaction it affords, pleasure in the possession of power, in "being a cause"—such, for example, as the child feels when he tears a paper into fragments.*

Lessing expressed it abstractly when he said that we become more intensely conscious of our reality by means of such strong excitations,† and animals may have the same feeling as an accompaniment of instinctive activity, and especially of playful experimentation. It may be lacking in the very earliest infancy, but the little polar bear that delightedly tore the paper bag to bits certainly felt the pleasure of "being a cause"—"in working his own sweet will," as Schiller has it in his Kunstlern. This principle is even more applicable to the examples which follow, relating as they do to more mature animals. Before going on, however, I wish to call attention to the absurd form this pleasure in being a cause sometimes takes even in rational beings How many of us want to scribble or whittle or do something with our hands all the time, to break a twig and chew it while we walk, to strike the snow off walls as we pass,

* Die Seele des Kindes, p. 456.

† Letter to Mendelssohn, February 2, 1757.

to kick a pebble before us, to step on all the acorns on the pavement, to drum on the window pane, to hit the wine glasses together, to roll up litle balls of bread, etc. Perhaps to this category, too, belongs that inexplicable piece of folly, of which we are all guilty, that when, for instance, a perfectly trustworthy person reads aloud a telegram "Can not come—Henry," we are never satisfied till we read it ourselves.

The case of animals is much like our own. The impulse to experiment continues into advanced age, and constantly tends to rise above its instinctive origin to freer, more individual activity, so that the fully developed animal probably also feels something of the pleasure in exercising power, in being a cause.

Beckmann says, in speaking of the raccoon: "The caged creature devised a thousand ways to relieve the tedium of his many idle hours. Now he would sit up in a corner and, with a most serious expression, busy himself with binding a piece of straw across his nose; now he played absorbedly with the toes of his hind feet, or made dashes for the end of his waving tail. Then, having packed a quantity of hay in his pouch, he lay on his back and tried to keep the mass in place by holding his tail tightly across it with his fore feet. Whenever he could get at masonry he gnawed the mortar and did incredible damage in a short time. Then he sits down, like Jeremiah before the ruins of Jerusalem, in the midst of his heap of rubbish, looks darkly about, and, exhausted with so much work, loosens his collar with his fore paws. After a long fit of sulks he can be restored to good humor at once by the sight of a full water bucket, and he will make any effort to get near it. Then he proceeds to test the depth of the water carefully, for when he is playing at washing things he wants to dip only his paws in the water, not at all liking to stand in it up to his neck. After satisfying himself on this point, he steps with evident delight into the wet element and feels about on the bottom for something to wash. An old pot handle, a bit of porcelain, a snail shell, are favourite objects for his purpose and are used over and over again. Now he spies an old bottle in the distance which appears to be greatly in need of washing. He reaches for it, but his chain is too short, so without hesitation he lies down as a monkey would do, gaining in that way the length of his body, and rolls the bottle toward him with his outstretched hind foot. The next moment we see him up on his hind legs, slowly waddling back to the water, the big bottle clasped in his fore paws and strained against his breast. If he is disturbed in his attempt he behaves like a self-willed, spoiled child, throwing himself on his back and clinging with all fours so tightly to his beloved bottle that he can be lifted by it. When he at last becomes tired of his work in the water, he fishes his plaything out, sits cross-legged and rocks to and fro, constantly fingering and boring into the narrow neck of the bottle."

This so-called washing seems to be characteristic of various kinds of bears as well. I myself have observed one instance, in the case of a polar bear that rolled an iron pot to and fro in his bath tub, taking it at last to a little trough of running water and there washing the indestructible vessel in earnest. It was very funny to see the bear seize it firmly with his fore paws and go through the motions of a washwoman scrubbing on a board. When the bath was freshly cemented in this bear house the animals were kept out of it for a day after the work was finished, for it was well known that they would soon spoil undried cement with their claws.

The following relates to dogs that have well passed their infancy. It may be called experimentation when a dog presses or rather scratches a beetle to death with his paw, as they are given to doing with a strange mixture of curiosity and aversion A St. Bernard, three and a half years old, that I formerly owned, spent many hours every day gnawing to bits any pieces of wood he could get hold of, usually from our wood pile unfortunately.

Alix tells of an Arabian dog that frequently amused itself playing with its own shadow on the wall. "Now straightening up his long ears, now turning them to right or left, now throwing them back, he produced in this way strange figures which appeared to amuse him greatly." *

A trustworthy person once told me of a dog that had played so much with the damper of a stove that he understood perfectly well how to turn it on. That he ever did so with the intention of raising the temperature seems to me a hazardous statement.

Some of the examples so far given relate to the destructive impulse, which is, however, only an extended kind of experimentation. Thus Scheitlin relates of an elephant: † "How amusingly that elephant in Cassel acted when his attendant forgot him in his stable! He went into the house, collected all the movables—tables, chairs, stools, pictures, and even the bed from the chamber—in a pile in the sitting-room, wet them all over, and walked out in the field as if he had not been at any mischief."

* E. Alix, L'esprit de nos bêtes, p. 440.

† Thierseelenkunde, ii, p. 178.

The destructiveness of monkeys is proverbial. They gnaw boards as dogs do—at least I have seen it done by the baboon and chimpanzee, their eating trough being badly disfigured in that way. Long-tailed monkeys amuse themselves by breaking off tough branches as they clamber from limb to limb.*

The book of daily observations, for which we are indebted to Romanes's sister, is full of examples of experimentation. It relates to a specimen of *Cebus fatuellus* which Romanes gave to his sister in December, 1880. The following description is from her diary:

"I notice that the love of mischief is very strong in him. To-day he got hold of a wineglass and an egg-cup. The glass he dashed on the floor with all his might, and of course broke it. Finding, however, that the egg-cup would not break for being thrown down, he looked round for some hard substance against which to dash it. The post of the brass bedstead appearing to be suitable for the purpose, he raised the egg-cup high above his head and gave it several hard blows When it was completely smashed he was quite satisfied He breaks a stick by passing it down between a heavy object and the wall, and then hanging on to the end, thus breaking it across the heavy object. He frequently destroys an article of dress by carefully pulling out the threads (thus unripping it) before he begins to tear it with his teeth in a violent manner.

"In accordance with his desire for mischief, he is, of course, very fond of upsetting things, but he always takes great care that they do not fall on himself. Thus he will pull a chair toward him till it is almost overbalanced, then he intently fixes his eyes on the top

* Loango Expedition, ii, p. 239.

bar of the back, and when he sees it coming over his way, darts from underneath and watches the fall with great delight; and similarly with heavier things. There is a washstand, for example, with a heavy marble top, which he has with great labour upset several times, but always without hurting himself." *

Vosmaern had a tame female orang-outang that could untie the most intricate knots with fingers or teeth, and took such pleasure in doing it that she regularly untied the shoes of those who came near her. Still more remarkable is the dexterity of Miss Romanes's monkey. Her entry for January 14, 1881, runs thus: "To-day he obtained possession of a hearth-brush, one of the kind which has the handle screwed into the brush. He soon found the way to unscrew the handle, and having done that he immediately began to try to find out the way to screw it in again. This he in time accomplished. At first he put the wrong end of the handle into the hole, but turned it round and round the right way of screwing. Finding it did not hold, he turned the other end of the handle and carefully stuck it into the hole, and began to turn it the right way. It was, of course, a very difficult feat for him to perform, for he required both his hands to hold the handle in the proper position and to turn it between his hands in order to screw it in, and the long bristles of the brush prevented it from remaining steady or with the right side up. He held the brush with his hind hand, but even so it was very difficult for him to get the first turn of the screw to fit into the thread; he worked at it, however, with the most unwearying perseverance until he got the first turn of the screw

* Romanes, Animal Intelligence, p. 484.

to catch, and he then quickly turned it round and round until it was screwed up to the end. The most remarkable thing was that, however often he was disappointed in the beginning, he never was induced to try turning the handle the wrong way; he always screwed it from right to left. As soon as he had accomplished his wish he unscrewed it again, and then screwed it in again, the second time rather more easily than the first, and so on many times. When he had become by practice tolerably perfect in screwing and unscrewing, he gave it up and took to some other amusement. One remarkable thing is that he should take so much trouble to do that which is of no material benefit to him. . . . It is not the desire of praise, as he never notices people looking on; it is simply the desire to achieve an object for the sake of achieving an object, and he never rests nor allows his attention to be distracted until it is done." * The report for February 10, 1881, runs: "We gave him a bundle of sticks this morning, and he amused himself all day by poking them in the fire and pulling them out again to smell the smoking end. He likewise pulls hot cinders from the grate and passes them over his head and chest, evidently enjoying the warmth, but never burning himself. He also puts hot ashes on his head. I gave him some paper, and, as he can not from the length of his chain quite reach the fire, he rolled the paper up into the form of a stick and then put it into the fire, pulling it out as soon as it caught light, and watching the blaze in the fender with great satisfaction. I gave him a whole newspaper, and he tore it in pieces, rolled up each piece, as I have described, to make it long

* Romanes, Animal Intelligence, p. 490.

enough to reach the fire, and so burned it all piece by piece."*

We here see the playful experimentation, which at first only serves the purpose of gaining control of the bodily organs, become further and further developed. No doubt, according to Darwin's theory of evolution, primitive man acquired the ability to use fire by just such experimentation.

The destructive impulse is manifested even more strongly by parrots and some other birds than by monkeys. Their winter quarters are often patched and mended like little Roland's cloak in Uhland's story, and the stronger the repairs the more eagerly do the parrots attack them. Linden tells of the persistence with which his cockatoos turned over the feeding trough in their cage. "I have used every device to make the troughs fast, winding fine wire about them and to the iron bars, screwing them tightly from the outside, etc., but my cockatoos know very well how to unscrew, and get them loose sooner or later." "The desire to do mischief is characteristic of the cockatoo," says Brehm, "and the performances of these birds pass belief. They gnaw through planks five or six centimetres thick, as I can testify from my own experience, and even iron plates a millimetre thick; they smash glass, and try to penetrate masonry." Rey relates of Carolina parrots: "Their favourite mischief was throwing their water-cups out of the cage after they had satisfied their thirst. Their delight was evident if the cups broke." And Dickens says, with delightful exaggeration, of a raven that died young: "It may have been that he was too bright a genius to live long, or it may

* Romanes, Animal Intelligence, p. 493.

have been that he took some pernicious substance into his bill, and thence into his maw, which is not improbable, seeing that he new-pointed the greater part of the garden wall by digging out the mortar, broke countless squares of glass by scraping away the putty all round the frames, and tore up and swallowed, in splinters, the greater part of a wooden staircase of six steps and a landing." * Brehm's brother had a tame vulture that often played with his master's fingers, taking them in his beak without hurting them. Another bird of the same kind, observed by Girtanner, tore the strong padding of his cage in every direction, took the straw out and played with it. He also clung to Girtanner's watch chain and clothing, "pulled straw from my hand, chuckling with delight. He took pleasure in tearing or biting straw ropes, and came to me at once when he saw that I was getting ready to make one." Still another stroked his master (Baldenstein) with his beak, or stuck it up his sleeve and uttered his contented "Gich"

Animals often amuse themselves by making noises. According to Scheitlin, hares can readily be trained to drum, because the motion is natural to them. "They drum with unexampled rapidity, quicker than any drummer boy, and even with a sort of passion." † This enjoyment of noise forms part of their pleasure in breaking and tearing. Experiments with apes especially illustrate this. Savage thinks that chimpanzees collect on purpose to play, on those occasions when they beat with rods on sounding pieces of wood ‡ This remark, in which I at first had little faith, has been fully

* Introduction to Barnaby Rudge.

† Thierseelenkunde, ii, p. 117.

‡ Romanes, Animal Intelligence, p. 476.

confirmed by the report of the Loango Expedition. Falkenstein tells there of a young gorilla: "A peculiar, almost childish, pleasure was awakened in him by striking on hollow, sounding bodies. He seldom missed the opportunity, in passing casks, dishes, or griddles, of drumming on them. On our homeward voyage he indulged freely in this pastime, being allowed to move about on the steamer." *

The same gorilla, too, frequently beat on his own breast with both fists, "apparently from overflowing contentment and sheer pleasure," a habit which in the adult usually indicates strong emotion, especially anger.

Voice practice is very common. I have already spoken of a puppy's attempts to bark, and I am inclined to think that even the howling of a young dog may be a kind of play; and I believe the same is true of young lions, that from time to time rise up and give forth frightful roars that commonly excite the others. The purring of cats, too, is like play. Then there are the deafening cries of the howling ape, considered by many as only an amusement. The wonder is that the animals have attained such a structure of the larynx as to be able to produce them. One kind of ape produces a flutelike note resembling the whistle of a bird, for which the lips are contracted. "Usually it is when he is unemployed, and seems to express his *ennui* by means of the sound." †

In many cases the vocal exercise consists in learning by heart a simple or complicated decoy cry that is usually connected with courtship, to which I will devote the next chapter. A single other example of voice practice will suffice, as it is a very valuable one. Hud-

* Loango Expedition, ii, p. 154.

† Rengger, Die Saugethiere von Paraguay, p. 45.

son relates of the crested screamer, or chakar (*Chauna chavarria*), that has a very loud voice: "There is something strangely impressive in these spontaneous outbursts of a melody so powerful from one of these large flocks, and, though accustomed to hear these birds from childhood, I have often been astonished at some new effect produced by a large multitude singing under certain conditions. Travelling alone one summer day, I came at noon to a lake on the pampas called Kakel, a sheet of water narrow enough for one to see across. Chakars in countless numbers were gathered along its shores, but they were all arranged in well-defined flocks, averaging about five hundred birds in each flock. These flocks seemed to extend all round the lake, and had probably been driven by the drought from all the plains around to this spot. Presently one flock near me began singing, and continued their powerful chant for three or four minutes; when they ceased, the next flock took up the strains, and after it the next, and so on until the notes of the flocks of the opposite shore came floating strong and clear across the water—then passed away, growing fainter and fainter, until once more the sound approached me, travelling round to my side again The effect was very curious, and I was astonished at the orderly way with which each flock waited its turn to sing, instead of a general outburst taking place after the first flock had given the signal. On another occasion I was still more impressed, for here the largest number of birds I have ever found congregated at one place sang all together. This was on the southern pampas, at a place called Gualicho, where I had ridden for an hour before sunset over a marshy plain where there was still much standing water in the rushy pools, though it was at the height of the dry season. This whole plain was covered with an endless flock of chakars, not in close order, but scattered about in pairs and small groups. In this desolate place I found a gaucho and his family, and I spent the night with them. . . . About nine o'clock we were eating supper in the rancho, when suddenly the entire multitude of birds covering the marsh for miles around burst forth into a tremendous evening song. It is impossible to describe the effect of this mighty rush of sound. . . . One peculiarity was that in this mighty noise, which sounded louder than the sea thundering on a rocky coast, I seemed to be able to distinguish hundreds, even thousands, of individual voices.

"Forgetting my supper, I sat motionless and overcome with astonishment, while the air and even the frail rancho seemed to be trembling in that tempest of sound.

"When it ceased, my host remarked, with a smile, 'We are accustomed to this, señor—every evening we have this concert.' It was a concert worth riding a hundred miles to hear." *

Much might be said of the twittering of sparrows, the quacking of ducks and geese, the flapping of storks, etc.; but, as has been remarked, it is difficult to determine how far such phenomena, especially the complicated ones, are connected with courtship. I reserve for the next chapter a closer examination of them. However, it may be noted here that in merely experimental noises and voice practice there is a suggestion of art which is not connected with courtship.

2. *Movement Plays.*

By this term I designate plays that involve change of place for its own sake. Hunting and fighting, in-

* W. H. Hudson, The Naturalist in La Plata, 1895, p. 227.

deed, also produce change of place to a considerable extent, but with them the movement has a specific aim. Here I refer only to such plays as are concerned with practice in locomotion as such, where the walking, running, leaping, climbing, flying, swimming of the animal finds its object in itself. As I said before, I pass by the lower orders, though some of their actions, especially the swarming of insects, is very suggestive of play. "With what joy in life insects swarm in the sunshine!" says Schiller; and Hudson is quite of the same opinion when he says: "I have spoken of the firefly's 'pastime' advisedly, for I have really never been able to detect it doing anything in the evening beyond flitting aimlessly about, like house flies in a room, hovering and revolving in company by the hour, apparently for amusement." * It may well be that animals quite low in the scale of being play, but who can prove it? "*Ludunt in aquis pisces*," says Julius Cæsar Bulengerus. Is it true that the fish tumble about so happily in their element? Is not this supposition rather the product of æsthetic sympathy—of the poetic delight that we ourselves experience on beholding the light, graceful movements of these delicate creatures? "In very large aquariums or in its native waters the stickleback swims along rapidly and gracefully, often leaping high out of the water, indulging in many gambols, but careful in it all to keep watch of what goes before it—namely, the young fry that forms its principle diet" (Brehm). How are we to know in such a case that all the movements do not serve the serious business of getting food? According to Noll, male and female carp chase one another playfully and delight in sportive leaping.

* The Naturalist in La Plata, p. 170.

But who can say that sexual instinct is not responsible for this? And the same may be said of the art of the flying fish. Brehm says of them: "On board ship, swarms of such fish can be seen at varying distances; they suddenly rise from the waves with a peculiar whirring and shoot rapidly over the water, sometimes rising to a height of four or five metres from the surface and travelling a hundred to a hundred and twenty metres before vanishing again in the waves. Not seldom this spectacle is quickly repeated, for as soon as one company rises, flies forward and falls, another begins to advance in the same way, and before it sinks a third and fourth are on the way. If these advances were made in a continuous direction we might suppose that their flight over the waves was to escape some danger. But they appear now here, now there, and keep to no particular direction, but fly across and contrary to one another. We can only suppose, therefore, that it is all a play, perhaps from pure exuberance of spirits, just as other fish swim rapidly over the water." Humboldt expresses the same view of flying fish. The sceptic may, of course, question whether all the motions described may not be attributed to flight or the search for food. Yet such an animal psychologist as Romanes speaks with great assurance of the play of fishes.* He says: "Nothing can well be more expressive of sportive glee than many of their movements." †

I am by no means so fully convinced as Romanes, but still I consider it highly probable that movement plays are manifested by fish. Their comparatively weak mental endowment is not a difficulty to me, since I re-

* Mental Evolution in Animals, p. 382.

† Animal Intelligence, p. 247.

gard play as at first an instinct, producing activity without serious motive. There can be no question that they often seem to play as they tumble about, and Romanes himself can offer no more convincing proof than that. The intelligence of fish is not, however, so inferior as is commonly supposed, and the probability that they have movement plays becomes apparent from the following observation of Beneke. He studied the habits of *Macropods* thoroughly, and made a report on them in Brehm's Thierleben, including a detailed account of the courtship of these fish: "The male usually, though not invariably, keeps to one particular female. On approaching her he extends his tail and fins, and grows perceptibly darker, while the female either remains perpendicular, all her fins closely compressed, and circles slowly round, or swims as the male does, though in the opposite direction. Then they turn slowly in circles together, the tail of one in front of the other's head, both with stiffly distended fins. If they become greatly excited during the play, the male trembles while he spreads himself, very much as a cock does when he struts around the hens, and the female often imitates this." When his male fish died Beneke secured another pair, and he says that the two females played together in the same way. The playful character of this can hardly be questioned, and, having admitted one case, we can not deny that much of the tumbling about in the water may really be playful. Of birds, however, we can speak with greater certainty. There are, it is true, many phenomena which have the appearance of play, but really belong to the search for food. Nothing, for instance, seems freer, lighter, or more aimless than the flight of swallows in spring, and yet we know that the impulse to satisfy their hunger, and not sportiveness, is the reason for it. The same is true, as a rule, of the cheerful hopping of birds from bough to bough and to the ground. Courtship, too, is at the bottom of much of the playful motion, as well as of voice practice. Referring this class to the next chapter, I here confine myself to a series of examples, most of which can be attributed with certainty to purely play impulse, and the remainder with great probability. First we notice the learning to fly, swim, or walk by young birds.

Birds can no more fly of themselves than babies can walk. The infant's kicking corresponds to the fluttering of little birds in the nest and his first step to its first attempt at flight. The tiny creature is very timid, and hardly dares to trust itself in the air. According to Hermann Müller's observation, a canary bird makes its first attempt to climb up on the nest rim on about the sixteenth day. Weinland gives a detailed account of a canary family: "Sixteenth day, 8 A. M.: The young dare not climb out of the nest, but reach and stretch a great deal. 10 A. M.: Amid great tumult one fluttered onto the rim of the nest and perched there, breathing hard and fast, appearing to be frightened at his own daring. In a minute the forward youngster is back in the nest. Seventeenth day, 7 A. M.: The feet as yet serve only as wide supports, like those of the ostrich, and not for dexterously clinging to boughs, as will be their later function. Twelve o'clock: Little Blackhead, the stronger one, has hopped out on the perch near the nest and down on the floor of the cage, and from there through the door of another cage, then quickly back. The little feet are still very unsteady, especially on the perch. On the ground he sometimes steadies himself with his tail, a use which is not made of it in later life. Eighteenth day: Both little ones have hopped about in the cage several times for some minutes and then back into the nest. Blackhead is always the leader. Twentieth day: Blackhead flew out of the cage; he found no place to light, fluttering high in the air all the time. He made the rounds of the ceiling several times, and at last, tired out, he fluttered down the wall to the floor. Twenty-first day: The yellow one also flies out in the room now. They can not find the way back yet. Twenty-third day: Blackhead took a bath. He plunged into the large shallow basin, made some awkward fluttering motions, and hurried out on the other side. Twenty-fourth day: Both birds fly, eat, bathe, and make their toilet alone." *

Dr. Krauss writes of the flying lessons of young storks: "As soon as the young can stand firmly and get to the edge of the nest, preparations for flight begin. Flapping their wings, they move round the nest, at first without rising above it. Then with a kind of hop they go a little higher, always hovering over the nest and keeping up this climbing process until they are at least a yard or two above it; they are able to continue the hovering a half minute or longer, after which they anxiously cling to the horizontal projection of the nest. Only when this has been repeated several times do they break the magic circle, gliding boldly out into the open air, describing in their flight a circle fifty or sixty metres in diameter, above the nest. They repeat this once, and then either fly back to the nest or rest on some neighbouring roof. At the end of July or the beginning of August begins the practice in high flying, preparatory to the great migration." †

The parents of sparrows, shy of flight, urge them on

* A Bird Family, Der Zoologische Garten, June. 1891.

† Krauss, Aus dem Freileben des Weissen Storchs, Der Zoologischen Garten, ix (1868), p. 131.

by holding food before them and flying on with the dainty morsel, uttering encouraging calls.*

"In the spring of 1872," writes Liebe to Brehm, "I noticed a pair of falcons circling over a wood. They were the terror of cranes living in that region. I happened to be passing there daily, and saw that for eight days one of the falcons came every evening to the wood and perched in a tree for about a quarter of an hour. After that he flew searchingly around the valley from time to time. I thought that the female must have been shot, but this suspicion was not confirmed. After some days she came again to the wood with the male at the usual hour, between 6 and 7 P. M., accompanied also by two young ones, which were still so helpless as hardly to be able to keep their equilibrium when perching in the trees. Soon both the old birds were skimming through the air, flying against the wind in their play. A beautiful performance, which I had once seen in Norway, and once by the male of this same pair. The male soon settled down, but the female kept up her wonderful evolutions, constantly drawing nearer to the young ones, till at last she shoved one of them, with a side push, from the bough, whether with her wings or breast my glass was not strong enough for me to see. The little one must fly whether willing or no, and after clumsily trying to imitate the movements of the old bird, it soon lit again. Thereupon the mother pushed the other one off its high perch and compelled it to fly like the first. Shortly they made both the young ones practice together, drove them aloft slantingly against the wind, shot perpendicularly down and then up again in splendid curves, and displayed all the skill that be-

---

* A. and K. Muller, Thiere der Heimath, i, p. 28.

longs to their kind. The little ones, trying to do the same things, awkwardly imitated their movements." Such actions are not rare in the animal world, where play and instruction are united, though in this case, with Brehm, too much is made of the analogy to human conduct.*

I class the learning to swim of aquatic birds among play movements. Here, too, the parents assist instinct, and so hasten their preparation for life's tasks. Old swimming birds take their young on their backs and then slide them off into the water—a very simple method, by which many a boy has been taught to swim. Julius Tapé gives a very beautiful description. He lived for a long time on the Danube, and "often noticed that young geese were afraid of the water before they learned to swim, and only gradually became accustomed to it by being, as it were, outwitted by the old ones. As soon as the little ones are old enough to go on the water their parents take them to the bank. The gander goes before, keeping up a continual gabble, and the mother, also gabbling, urges them on from behind. After a very short trial of swimming the young ones are quickly brought back to land, and this trial is repeated and lengthened from day to day until they go into the water alone." † That swimming is not entirely an acquired art, however, but instinctive in part, is proved by the case of ducks hatched by a hen. How Büchner can find an argument against instinct in the fact that these little ducks need a longer time to become accustomed to the water I can not see. Hermann Muller

* So in the teaching of young beasts of prey to seize their victim.

† L. Büchner, Aus dem Geistesleben der Thiere, p. 31.

says of young birds learning to walk: "The first movements seem to be not on the toes, but on the heels. If they are hurried, the birds tip forward, steadying themselves with their wings." Buchner describes the walking of little chickens, from Stiebeling's observations: "The chicken begins, probably about two hours after it breaks the shell, to make feeble attempts at walking, in which the wings serve as crutches. He rises and sinks again, falls down and gets up again, so that the whole process is more a slide than a walk. It learns to walk in from five to eight hours if its mother helps it, but from eight to sixteen hours are needed if the chicks are separated from the hen as soon as hatched." *

Such movements can of course be considered as play only so long as they are simply exercise. As soon as the bird is far enough on to turn his flying to account in the search for food, play changes to serious activity. This transition takes place very quickly in birds, but their short time for practice is just as really a playtime as is the longer period of beasts of prey.

Some phenomena belonging to migration ought perhaps to be mentioned in this connection. That this impulse is instinctive is witnessed to by the classic ornithologist Naumann, in a passage already quoted. "The impulse to seek a warmer climate," he says, "is hereditary in these birds. Young birds taken from the nest and placed in a large room, where they are allowed to fly about freely, prove this conclusively. They are restless at night during the season for their migration, just

---

* Ibid., p. 31. From Stiebeling's Instinct in Chickens and Ducks, New York, 1872.

as old birds of their kind are." * Before the time for their departure migratory birds are fond of collecting in large flocks, and this can only be regarded as play, especially in the case of the young, preparatory for the long flight. Thus, in the spring, young nightingales take little experimental trips from shrub to shrub and field to field. It is the same with the young of the whitethroat, bower bird, song thrush, and many other kinds of birds.† Though it is doubtful, as I have said, whether the so-called flying games of adult birds are movement plays, I will include a couple of such examples. Scheitlin tells of a young crane: "He went to the field with his master, rose in the air of his own accord and with evident pleasure, tumbled about some, and then came down and walked by his master's side."

Hudson relates of the wonderful crested screamer: "I was once very much surprised at the behaviour of a couple of chakars during a thunderstorm. On a sultry day in summer I was standing watching masses of black cloud coming rapidly over the sky, while a hundred yards from me stood the two birds, also apparently watching the approaching storm with interest. Presently the edge of the cloud touched the sun and a twilight gloom fell on the earth. The very moment the sun disappeared, the birds rose up and soon began singing their long-sounding notes, though it was loudly thundering at the time, while vivid flashes of lightning lit the black cloud overhead at short intervals I watched their flight and listened to their notes, till suddenly, as they made a wide sweep upward, they disappeared in

* J. A. Naumann, Naturgeschichte der Vögel Deutschlands, i, p. 86.

† A. and K. Müller, Thiere der Heimath, i, p. 81.

the cloud, and at the same moment their voices became muffled, and seemed to come from an immense distance. The cloud continued emitting sharp flashes of lightning, but the birds never reappeared, and after six or seven minutes once more their notes sounded clear and loud above the muttering thunder. I suppose they had passed through the cloud into the clear atmosphere above it, but I was extremely surprised at their fearlessness." The beautiful floating motions of birds of prey are principally for reconnoitring, and are also connected with courtship, but it may well be supposed that the birds sometimes exercise their skill from pure pleasure in the movement. Darwin tells us that the condor gives a splendid exhibition, floating for half an hour without a movement of the wings, describing great circles, rising and falling in beautiful curves.

What has been said with regard to the art of flying applies also to the dancing of many birds, except that I consider the connection with sexual instinct closer in the latter case, where many of the movements are highly specialized. Hudson, on the contrary, regards the dances of birds as purely playful, originating in cheerful spirits. Although I do not agree with him, I must admit that the sexual explanation is impossible in the case of one of his examples. He is speaking of the spur-winged lapwing, that resembles the European lapwing, but is a third larger, more highly coloured, and furnished with spurs on its wings. Three individuals are required to perform their dance, which, according to Hudson, is unique in this respect. "The birds are so fond of it that they indulge in it all the year round and at frequent intervals during the day, also on moonlight nights. If a person watches any two birds for some time—for they live in pairs—he will see another lapwing, one of a neighbouring couple, rise up and fly to them, . . . and is welcomed with notes and signs of pleasure. Advancing to the visitor, they place themselves behind it; then all three, keeping step, begin a rapid march, uttering resonant drumming notes in time with their movements. . . . The march ceases; the leader elevates his wings and stands motionless and erect, still uttering loud notes; while the other two, with puffed-out plumage and standing exactly abreast, stoop forward and downward until the tips of their beaks touch the ground, and, sinking their rhythmical voices to a murmur, remain for some time in this posture. The performance is then over, and the visitor goes back to his own ground and mate to receive a visitor himself later on." * If this description is entirely accurate, the foregoing will probably long remain one of the unsolved riddles of animal life.

Finally, the swinging that gives such pleasure to many birds must be included in the list of movement plays. Every one knows how captive parrots and canaries love to swing on a ring, and it appears from the observation of Naumann that birds often cling to the highest tip of a swaying bough to swing on it. He has seen the blue titmouse, the bearded titmouse, penduline titmouse, thistle finch, barley bird, birch siskin, and others do this.†

But I must now leave the interesting world of birds and turn to some other phenomena. Finsch has observed the habits of seals in the vicinity of San Francisco, and describes them graphically. While the move-

---

* The Naturalist in La Plata. p. 269.

† J. A. Naumann, Naturgeschichte der Vögel Deutschlands, iv, pp. 67, 68, 88, 107, 120; v, pp. 134, 163, 182.

ments of these lumbering animals on land are remarkable, it is in the water that their skill is most admirably displayed. They may be seen plunging into the sea, either sliding down the smooth, sloping sand banks or throwing themselves from a high rock; then they carry on their play like dolphins, rapidly throwing themselves over so that the belly is uppermost, and sometimes springing entirely out of the water. They swim round in circles, now and then leaping up, splash about, whirling and turning and tumbling about like mad, and so entirely forgetting themselves that the wary hunter can easily come within harpooning distance and capture them. The behaviour of seals in captivity is equally remarkable.

In speaking of dolphins, Lösche says: "Every seaman is delighted to see a school of dolphins. The cheery travellers hurry along through the swelling waves in a long and regular train, pursuing their way with a speed that suggests a race, and with leaps of wonderful agility. Their glittering bodies rise in the air in fine curves from one to two yards wide, fall headlong into the water, and soon spring up again, carrying on the game. The jolliest of the flock turn somersaults in the air, turning up their tails in a most comical manner. Others fall flat on side or back, and still others remain bolt upright, dancing along with the help of their tails until they have made three or four forward movements. They no sooner see a ship under full sail than they turn about and follow it, and then begins real sport. They circle around the vessel, leap in front of it, and make the best possible exhibition of their skill. The faster the boat moves the more riotous their antics." Brehm gives this description of the exercise of a caged marten: "He amused himself for hours at a time making bounds that brought him to one wall of his cage, where he quickly turned and sprang back, landing in the middle of the floor, then to the other wall and back—in short, describing the figure 8, and with such rapidity that its outline seemed to be formed of the animal's body." *

A caged fox that I have observed behaved in the same way, except that his motion was in a circle, because on leaping to one wall he rebounded to the opposite one, and only then came to the floor. Every visitor to zoological gardens or menageries is familiar with the tiger's ceaseless walking up and down, the constant waving to and fro of the badger's and bear's fore paws, and other such motions. They are all playful, and are the best possible examples of discharge of superabundant nerve force; for, of course, caged animals do not have a sufficient outlet for their energies. However, the kind of movement is not determined by outward circumstances, but, like all play, rests on an instinctive basis. A hunter, cited by Tschudi, testifies that the badger when wild and free and especially comfortable waves his fore paws indolently. The decidedly rhythmical character of such movements is noteworthy. Indeed, they tend to prove that all free motion unimpeded by other forces is likely to be rhythmical.

Schlegel tells of a tame leopard that was very fond of children—"especially of a little girl five years old, whom he often jumped over in play, and with such ease that without any preparatory running he crouched and easily bounded higher than the child's head."

* When free, too, the young marten is much inclined to movement play, restlessly busying himself with all sorts of climbing and leaping exercises (A. and K. Müller, Höhern Thierwelt, p. 75).

Young bears are exceedingly playful. One that I watched for a long time galloped with indefatigable energy around the great kennel, directing his course through the water pool each time. His noisy splashing seemed to give him particular pleasure. The young badgers in Regent's Park, London, amuse visitors by turning somersaults hundreds of times in succession in the same spot. The wild buck gives expression to its joy in graceful, sportive leaps.* Such leaps, alternating with tearing madly around, are expressions of well-being which so intoxicate the young hare that his worst enemy, the fox, creeps up unawares. Buffaloes, tapirs, and crocodiles sport in the water as night comes on. The leaping of young horses, asses, sheep, and goats is familiar. A phenomenon pointed out to me by Director Seitz illustrates how closely such movement plays are connected with habits which are indispensable in the serious struggle for life. He writes: "It is my impression that, in general, the play of animals exercises them in directions that will be useful for them in the necessary struggle for existence. The gazelle practises long jumping and leaping over bushes; goats and sheep, that live in mountains, the direct high jump." Many will be surprised to find an explanation for such goat leaps, which usually make us laugh, and are certainly extraordinary movements and wholly inexplicable on level ground. They are, however, necessary practice for life in rocky hills.

"A two-weeks-old goat," says Lenz, "not satisfied with the remarkable leaping record which he had already made, had the greatest desire to attempt breakneck feats. His motto was 'Excelsior.' His greatest

* A. and K. Müller, Thiere der Heimath, i, p. 422.

pleasure was to clamber on piles of wood or stone, on walls and rocks, and to mount the stairs." *

The purely playful motions of cats should be mentioned here. They delight in racing about, but not so often, I think, in circles, as dogs do. They prefer straight lines and sharp turns with the genuine goat jump. This sudden flight into the air, which appears to take place without the animal's knowledge or intention, can not here be preparatory to life in the mountains, but the cat finds the high jump very useful, not only in pouncing on its prey, but also in escaping its hereditary enemy. Chamois are, of course, adepts in high jumping. A very remarkable movement play is reported of them, whose actual occurrence was vouched for to Brehm by two witnesses. "When in summer the chamois climb up to the perpetual snow, they delight to play on it. They throw themselves in a crouching position on the upper end of a steep, snow-covered incline, work all four legs with a swimming motion to get a start, and then slide down on the surface of the snow, often traversing a distance of from a hundred to a hundred and fifty metres in this way, while the snow flies up and covers them with a fine powder. Arrived at the bottom, they spring to their feet and slowly clamber up again the distance they have slidden down. The rest of the flock watch their sliding comrades approvingly, and one by one begin the same game. Often a chamois travels down the snow slide two or three times, or even more. Several of them frequently come roughly together at the bottom." If this description is to be relied upon, we have here, as in the swinging of birds and many other forms of experimentation,

* H. O. Lenz, Gemeinnützige Naturgeschichte, 1851, i, p. 612.

genuine play. I do not consider this coasting impossible, since the chamois must frequently make their way across snow-fields, and no doubt often slide down unintentionally. I have seen a young dog slide all the way across the room with his fore feet in a slipper, using his hind feet as propellers, and all the while snapping and snarling. In such cases accidental movements are made, which may be repeated intentionally later. The following incident, related by Alix, points more directly to this supposition than the account of the coasting chamois: "While manœuvring in the Alps with a squadron of my regiment, I was botanizing one day in the neighbourhood of Briançon, followed by one of those stray dogs that so frequently attach themselves to moving troops. Just as I was about to begin the descent by the interminable winding path which gives access to the defile, I noticed that the dog, instead of following me, turned toward one of the steep declivities of the mountain side, where there was an accumulation of snow. Being puzzled to understand his behaviour, I stood still and took in every movement of the animal. And I was well rewarded, for by so doing I became witness to a strange spectacle, most wonderful even to the man accustomed to the unlimited resourcefulness of dogs. Placing himself on his back, his paws folded, his head bent forward, the intelligent animal slid down on the snow crust to the very base of the mountain. Arrived at the edge of the snow bed, he quietly rose, cast his eye toward me, wagged his tail, and lay down on the grass to wait for me."

Alix supposes the dog to have reasoned that the way could thus be shortened. I consider more probable the rather vulgar explanation that the dog had learned this remarkable trick from rubbing his too populous back on the snow. However, this forms a companion piece to the tale of the chamois.

The effort of puppies to walk is the first manifestation of movement play. At first they can only creep about with difficulty, and when they learn to stand up, an attempt to bark is enough to upset them. As soon as they can stand decently they at once try to gallop, usually in a slanting direction. By constant practice the necessary accuracy is gained for carrying on their chasing and fighting games.

The play of grown dogs in water is noteworthy. The Newfoundland especially is such an enthusiastic swimmer that he has been known to leap from a bridge to get to his beloved element. However, as most of the play of dogs belongs in another category I shall not dwell on it here, except to record what in our family we call the run-fever, the aimless and objectless running about that is to be observed of little dogs in a large room, but of large dogs only in the open air. He tears about wildly, mostly in curves, though our fox terrier loves to dash off straight as a line to a great distance till he is lost to the eye of his vainly whistling master. It might be said that this points to imaginary prey,* and that this is accordingly a chase play rather than a movement play Romanes tells of a poodle, named Watch, that belonged to the Archbishop of Canterbury, that hunted for imaginary pigs when he heard the word called out. He went so far as to beg to be let out, running to the door for the purpose, and rushing out without any further instigation than that the word "pigs" should be mentioned. It is difficult to determine whether he really imagined the pigs or not, but such

* Or perhaps in some cases an imaginary flock to collect.

actions are common enough. For instance, my pug, who is a sworn foe of cats, flies to the garden and all along the fences if he hears the cry "St! cats!" I am doubtful, however, whether this is properly called play; at any rate, it is quite different from the run-fever, for now the pug runs with loud cries and sharp attention, while in the run-fever the dog moves off silently and looks neither to the right hand nor to the left. Consequently, I look upon the latter as play purely for the sake of the movement. Perhaps in a sense the same may be said of the propensity some dogs have for taking walks. A bulldog of very philosophical disposition that I owned when I lived in Heidelberg, took regular, solitary walks that threatened to be expensive to his master. He would go off without his muzzle, a thing forbidden by the authorities; could be seen strolling boldly past the police office, climbing the Schlossberg, and entering the garden of the palace, where dogs are not allowed unless led by some one Of course, we do not know how much weight to attribute to the attractions of digging under the curbing, sniffing at corners, and other pleasures of freedom, yet I am sure the dog delighted in the walk for its own sake, and am not afraid of contradiction on this point from those who know dogs.

Last of all we must consider the monkeys. Their movement plays may be divided into four groups: climbing, leaping, swinging, and dancing It is unnecessary to describe the behaviour of caged monkeys, for even the most careless sightseer stands long in front of a monkey house in a zoological garden. I therefore confine myself to some reports of their play when at liberty, supposing the clambering about a ship to be free motion. Captain Smith had an orang-outang three months on board his vessel and allowed him perfect freedom. Climbing and exercising in the rigging seemed to give him the greatest pleasure, for he indulged in it many times a day, and in such a manner as to astonish all beholders with his dexterity. Bennett makes a similar report of a specimen that he brought to Europe. A female ape (Spinnenaffe), whose habits have been well described by an Englishman, also played in the rigging. When "Sally" really wanted to have some fun she danced with such gaiety and abandon on the sail yards that a spectator could hardly distinguish arms, legs, and tail. At such times the name "spider ape" seemed especially appropriate, for she resembled a giant tarantula in her motions. During this performance she would stop from time to time and gaze, with familiar nods, at her admirers, wrinkle up her nose, and emit short low sounds. She was usually liveliest at about sunset. Her special delight was to clamber up the rigging till she reached a horizontal sail yard or a slender spar. Here she hung firmly by the end of her tail and swung to and fro.* According to Duvaucel, "the gibbon climbs with incredible rapidity up a bamboo stalk or to the top of a tree, there swings to and fro several times, and then, aided by the impetus so gained, throws himself a distance of twelve or thirteen metres. Repeating this three or four times in rapid succession, his progress appears much like the flight of a bird. One is forced to believe that the consciousness of his unparalleled dexterity affords him pleasure, for he leaps unnecessarily over spaces that he could easily avoid by a slight detour, varies his course, leaping to promising boughs, swinging and hanging there, and again launching out in the air with unfailing certainty toward a new

* [See other cases in the original.]

goal. He seems to proceed magically, flying without wings; he lives more in the air than on the branches." *

The young gorilla of which J. Falkenstein gives so interesting a description, "performs so abandoned a dance, falling over himself, whirling about, tumbling from side to side, that the looker-on is forced to believe that he has in some way become intoxicated. And in truth he is drunk with pleasure, and by means of these antics he proves his own strength to himself" † The swinging of monkeys is also a proof of the invention of plays in the animal world. The explanation is not difficult, seeing that the movements are often made intentionally as the monkeys go about in the trees. The pleasure they take in it seems to be unlimited. Pechuel-Loesche tells us of one very clever ape that made himself a swing, a case that would have surprised Descartes! A tame long-tailed monkey that the members of the Loango Expedition kept at their station, a so-called Mbukubuku, "was a devotee of swinging to an unprecedented degree, and knew well how to satisfy his propensity. On any tree that he could reach, on the roof, and on his own kennel he found projections that served as supports, to which he fastened his long chain by climbing over them or going round in such a way that it caught, and in this way swung to his heart's content. He would go to work with admirable deliberation and measure off a length of his line sufficient for the purpose, and would repeat a successful manner of fastening even after months." ‡

* See Alix's description of a gibbon, L'esprit de nos bêtes, p. 496.

† Loango Expedition, ii, p. 152.

‡ Ibid., iii, p. 243.

3. *Hunting Plays.*

Instinct is much more conspicuous in this class of plays than in those which we have heretofore considered, for by means of them the young animal, even while yet having its food provided by parental care, practises sportively those movements which will be used in earnest later on.

Even the domestic animals—the dog for instance, that may never feed on prey, but eat all its life from the prosaic feeding trough—carry on with passionate zeal, plays the origin of which must be sought in the ancestral manner of feeding. A glance over this class of plays shows us that they naturally fall into three groups: (*a*) Play with actual living prey. (*b*) Play with living mock prey. Animals of the same kind usually chase one another reciprocally; thus we have to consider the letting themselves be chased, as well as the active chasing. (*c*) Play with lifeless mock prey, with a stick of wood, a ball, or other such objects. I have arranged the order of these groups so that the examples most illustrative of simple play shall come first, but it would be a mistake to suppose that actual time sequence is indicated by this order. On the contrary, play with lifeless objects is in many cases first in point of time.

(*a*) Is the treatment of living prey by carnivorous animals properly called play? A beast of prey seizes his victim, does not kill it, but lets the slightly wounded creature loose on the ground. It takes to flight, but is instantly recaptured, perhaps shaken a little, and again set free. This time it lies motionless, perhaps from weakness, perhaps to feign death. But the merciless beast keeps teasing it until it again attempts flight, only to be seized once more by its tormentor. In this way the "play" goes on until the victim really dies and is devoured. I was formerly of the opinion that the instinct here called out should not be regarded as play at all, but had an entirely different meaning. The explanation once suggested by G. Jaeger—namely, that it was done for the purpose of improving the flavor (as connoisseurs think that hunted game is especially good)—does not appear probable, though it can hardly be proved to be impossible. There may be some other reason unknown to us for this phenomenon which excludes it from the category of plays, but it is generally regarded as belonging there. Darwin unhesitatingly enumerates it among other plays,* and Scheitlin says of the cat: "She lets the mouse loose again and again in order to catch it each time, and plays with it unmercifully. Mouse and rolling ball are all alike to her, as the real and the toy beetle are to the child." † Even if this be true, there is still a difficulty. Granting that the animal sees no difference between the living and the lifeless, how are we to explain the awakening of playfulness in the very act of slaying, and so strongly as to hold in check that instinct, which is so powerful in a beast of prey?

The whole thing is usually ascribed to a natural instinct for cruelty. Even Romanes says, "The feelings that prompt a cat to torture a captured mouse can only, I think, be assigned to the category to which by common consent they are ascribed—delight in torturing for torture's sake."‡

If this is true it is undoubtedly a play. The disposition to cruelty would explain the tendency to play at

* The Descent of Man, ii, p. 52.

† Thierseelenkunde, ii, p. 222.

‡ Animal Intelligence, p. 413.

killing, even in the midst of the actual thing. But is this *consensus omnium* to be depended on? Is not pleasure from cruelty a kind of degenerate æsthetic satisfaction that requires higher intellectual capacity than animals possess? I do not venture an assertion, but I confess that the current idea seems to me very improbable. A remark of Dr. Seitz in a letter to me seems to be nearer the truth: "The cat's play with a captive mouse probably serves to practise the springing movements, as well as affords an opportunity to study the mouse's way of running and to acquire the necessary stealth in ambush."

Thus torture of living prey would be an instinctive exercise for acquiring skill in the chase, later turned to account by the animal; it is a play, whose usefulness accounts for its existence, unusual as it is. Appearing in early youth, it becomes firmly established in riper years, and the pleasure in being a cause plays its part.

Without assuming a positive attitude on this question, I proceed to adduce some examples of torture by beasts of prey. The cat has been referred to, and every one is familiar with its habits The wild cat also, according to Scheitlin, plays with captive mice and birds. A leopard that belonged to Raffles played for hours with the fowls that were fed to him on the ship. Indeed, it can safely be said that such behaviour is characteristic of the whole feline tribe.

"Most of the cat family," says Brehm, "have the horrible practice of torturing their victims, pretending to set them at liberty, until the wretched creatures at last succumb to their wounds." Lenz relates of a marten: "His hunger satisfied, he would play for hours with the birds brought to him. He liked little marmots even better; he leaped and sprang about the incensed and spitting animals, incessantly dealing them blows first with the right paw and then with the left. If he were hungry, however, he made no such delay, but devoured them at once, bones, skin, and hair." *

I have included the dog's toying with a beetle under the head of experimentation, though perhaps it would be more appropriately placed here, for my terrier plays with mice that he catches just as a cat does. It is certain, too, that foxes torment their victims long and cruelly and instruct their young in the art.† The mother weasel brings living mice to her little ones to play with and to practise on.‡

"In Altures," says Humboldt, "we had an adventure with a jaguar. Two children, a boy and a girl of eight and nine years, were playing near the village. A jaguar came out of the woods and bounded near them. After leaping about for some time, he struck the boy on the head with his paw, at first softly, and then so hard that the blood streamed forth. At this the little girl seized a stick and beat the animal till it fled. The jaguar seemed to be playing with the children, as a cat does with mice." # Finally, I may mention the cormorant that is described in Darwin's Descent of Man (ii, p. 52) as playing in a similar way with fishes.

(*b*) Play with living mock prey. An animal will play with another, usually but not always of the same kind, as he does with his prey. In that case both are playing, and the value of such practice for the serious tasks of after-life is evident. Among beasts of prey

* H. O. Lenz. Gemeinnützige Naturgesch., i, p 166.

† Ibid., i, p. 266.

‡ Müller, Thiere der Heimath, i, pp. 352, 355.

H. O. Lenz, Gemeinnützige Naturgesch., i, p. 327.

the pursuer is far more active and interested in the game than the fleeing one, while with herbivorous animals the contrary is the case; with these, as Dr. Seitz writes, the animal that is fleeing plays the principal part, the other merely co-operating and doing its share in a perfunctory sort of way. The dog offers an excellent example of the first class. A dog that sees another approaching, frequently crouches in the open street and remains quite motionless, with all the signs of eager alertness. This instinctive lying in wait is evidently rudimentary, for when the other dog comes up the one in ambush rises forthwith and goes to meet his comrade. Sometimes the dog goes so far as to hide himself. Not long ago I saw a young fox terrier leaping around the corner of a house to hide himself from another dog that was coming. Then followed the invitation to play, made in a very characteristic manner, with legs wide apart, a position well adapted to facilitate the rapid projection of the body in flight. All ready to start, he throws himself from right to left several times, in a semicircle, before the flight really begins. The other in the meantime is a fine picture of hypocrisy, as he glances indifferently about as if the whole affair were nothing to him. Now, however, the fun begins, as the leader springs forward, though not at full speed, and the other gives chase with enthusiasm. Should the pursuer overtake his mock prey, he tries to seize him in the neck or by the hind leg, just as a dog does when chasing in earnest. The other, without slackening his pace, turns his head to defend himself by biting. Then a tussle often ensues. At last the players stand with tongues hanging out, breathing heavily, until one of them suddenly whirls around and the play begins anew. The elements involved in all this are lying in wait, hiding, invitation to play, deception, fleeing, pursuing, overtaking, seizing, and defence. I am anxious to emphasize the various movements involved on account of the unsatisfactory nature of the reports of the series of examples I am about to relate. In order to avoid having too many subdivisions, I cite cases of play between animals of different kinds and between animals and men promiscuously.

"While I was staying in Tunis," says Alix, "my dog Sfax doted on playing hide and seek with the native babies. . . . Concealing himself among the wood-piles, Sfax described the most complicated zigzag, and just when five or six of the youngsters thought they were going to put their hands on him he would appear on a pile twenty metres away, sometimes in front, sometimes behind, sometimes to the right or left. He would stand there with an air of careless indifference till his playfellows ran within two or three metres of him; then gaily wagging his tail, he set off to make more zigzags, and so on for more than an hour." *

Young horses gallop about the meadows, leaping with joy; those grazing on the Russian steppes sometimes accompany travelling carriages at a gallop for many hours.†

Brehm says: "The tame cougar (puma) plays with his master, delighting to hide at his approach and then spring out unexpectedly, just as a tame lion does. It may well be supposed that such savage demonstration of affection is anything but agreeable at an inopportune moment." ‡ Hudson considers the puma the most

* L'esprit de nos bêtes, p. 498.

† Scheitlin, Thierseelenkunde, ii, p. 242.

‡ See Rengger, Saugethiere von Paraguay, p. 189. A tame lynx, brought up by O. v. Loewis, behaved in the same way. Der zoologische Garten, 1866, No. 4.

playful of animals, with the exception of the monkey. An Englishman told him the following incident: He was once compelled to spend the night in the open air in the pampas of La Plata. It was a bright moonlight night, and at about nine o'clock he saw four pumas approaching, two adult animals and two half-grown young ones. Knowing that these animals never attack men, he quietly watched them. After a while they came very near him as they chased one another and played at hide and seek like kittens; and finally they jumped directly over the motionless man several times. The mother cat will run forward some distance and call the little ones after her. P. Kropotkine had a cat that played regular hide and seek with him.* Monkeys do the same, both on the ground and among the branches.

Young wolves play just as dogs do, and it is at least in part a chase that Brehm describes of the weasel: "Until these charming little creatures are quite grown they play often during the day with their parents, and it is a sight as strange as it is beautiful to see a company of them collected in a meadow on a bright day. The play goes merrily on. From this or that hole a little head pops out and small bright eyes glance from side to side. Everything being quiet and safe, one after another comes out of the ground to the fresh grass. The brothers and sisters tease one another, romp and chase, and so cultivate the agility that is their natural inheritance."

The head forester Nördlinger relates the following of two ravens and a weasel: The latter had taken refuge in a street gutter. As quick as lightning he darted out, rustled through the dry leaves that partly covered the

* Revue scientifique, August 9, 1884.

ground, and made a pretended attack on one of the ravens; throwing himself about among the leaves, like a fish on land, and pretending to snatch the birds, he frightened his victims by the wildest, most dexterous leaps, during which the white belly was as often uppermost as the brown back. Then he fled back to the gutter, from which only his fore legs protruded; or he took up his position on the street, awaiting the attack of the raven which followed his own, and evidently with as little serious intent. The raven, with head outstretched, ran after the alert creature, but with small success, for he was not inclined to test his agility seriously against the powerful beak of one or perhaps both of the birds. The game lasted with many variations on both sides for about ten minutes, when it was interrupted by my dog, and the ravens flew away *

Beckmann very beautifully describes the play of a badger. "His only playmate was an exceedingly clever and sensible dog, which I had accustomed from its youth to live with all sorts of wild animals. Together they went through a series of gymnastic exercises on pleasant afternoons, and their four-footed friends came from far and near to witness the performance. The essentials of the game were that the badger, roaring and shaking his head like a wild boar, should charge upon the dog, as it stood about fifteen paces off, and strike him in the side with its head; the dog, leaping dexterously entirely over the badger, awaited a second and third attack, and then made his antagonist chase him all round the garden. If the badger managed to snap the dog's hind quarters an angry tussle ensued, but

* Müller, Thiere der Heimath, i, p. 351. See also Hudson, *loc. cit.*, p. 385.

never resulted in a real fight. If Caspar, the badger, lost his temper he drew off without turning round, and got up snorting and shaking and with bristling hair, and strutted about like an inflated turkeycock. After a few moments his hair would smooth down, and with some head-shaking and good-natured grunts the mad play would begin again."

Alix says that goats often play at hide and seek with the village children.* Young foxes play this game together, and so do squirrels.† The female marten carries on all sorts of gambolings with her young. The little ones run after her, she leaps over them, springs and whirls about like mad in every direction.‡ Fraulein Minna Haass, of Rosterberg, had a tame fawn named Lieschen that followed her mistress all about, came at her call, and manifested a real attachment for her. "The animal also cherished a friendship with two huge mastiffs, and delighted to play with them. When ready for a game, the fawn would approach the dogs as they lay before the door, tap them with her fore-foot, and take to flight. At this signal a game followed exactly like the hide and seek played by children, and a beautiful sight it was. If the dogs were disinclined to play, Lieschen kept urging them till they came." #

Antelopes when followed keep the same distance from a pursuer, as if they were mocking at him. Seals chase one another vigorously in the water. Birds, too, have a kind of play that is like chasing. Naumann says

* *Loc. cit.*. p. 173.

† A. and K. Müller. Wohnungen, Leben und Eigent. in der höheren Thierwelt, pp. 90 und 161.

‡ A. and K. Müller. Thiere der Heimath, vol. i. p. 364.

Büchner, Liebe und Liebesleben in der Thierwelt, p 263.

that, as autumn comes on, the redstart and their young may be seen chasing and teasing one another. Scheitlin tells of a tame stork "that very easily made friends, especially with children, and would even play with them, running after them with outstretched wings and catching hold of their coats or sleeves with his bill, and then back, looking round to see if the children followed. It would wait to be caught by the wing and then start after the children again. This scene was repeated as often as the children played 'catcher' in the street." * A. Gunzel relates of a tame and trained magpie. "At the time of the morning recess she went to the playground of the school children, especially of the boys, to look on while they romped. She expressed her pleasure by hopping about excitedly and snapping her bill. The boys loved to tease her. She would stretch her long tail out, and when any one tried to touch it, spring so nimbly to one side that they never succeeded in catching her. Even I could not touch her then, though at other times she was quite docile. She enjoyed this play, and would follow any one who caught at her tail in order to repeat the game." † The older Brehm relates of the golden-crested wren: "This little bird carries on a strange performance in the fall, from the beginning of September to the end of November. It begins by calling out repeatedly 'Si, si!' whirls around, and flaps its wings. Others answer to the call, and they collect, all going through the same motions. From two to six usually play together." ‡

"The woodpecker," writes Walter, "is an enthusiastic player, and often has his parents as playmates. A

* Naumann, iii, p. 531.

† Die gefiederte Welt, 1887.

‡ Chr. L. Brehm, Beiträge zur Vögelkunde, ii, p 126.

shaking twig or bit of cloth sets the whole family into the most joyous agitation for fully five minutes. They clamber about a tree like monkeys, hiding with outstretched wings behind the trunk till they are found, and then they all run and dance around the tree, chasing and teasing each other."

We must always remember in estimating these actions of birds that most of them are probably connected with courtship. But Huber's observations of ants —which, however, have been questioned—indicate that these insects actually do play at hiding and chasing *

(*c*) Play with lifeless objects. It usually appears, as I have said, before the other two kinds of play already mentioned.

The sportiveness of kittens is alone sufficient to prove that play is founded on instinct. The tiny creature creeps from its nest, still blind, but as soon as even one eye is open it toys with every rolling, running, sliding, or fluttering object in its reach,† and only when it has practised on such things and become prepared for the real business of a preying animal does the old cat bring living prey to it. In this case play is surely not the child of work, as Wundt calls it, but rather it is Ziegler who is right when he says that work is the child of play.‡ Various kinds of movements are distinguishable in a cat's play with balls, suspended cords, bundles of paper, etc. A moving object is best to test this with, for, "*cæteris paribus*, objects moving slowly fasten the attention most readily" #—a fact of significance in the struggle for life.

---

* See Büchner, Geistesleben der Thiere, p. 196.

† Scheitlin, Thierseelenkunde, ii, p. 217.

‡ Th. Ziegler, Das Gefühl, Stüttgart, 1893, p. 235.

L. William Stern, Zeit. fur Psych. u. Phys der Sinnes-

Observation of this motion produces in the young animal first perfect motionlessness, attended by that strenuous attention that we call "lying in wait," whose analogue is found in the feigning of death by an animal when pursued. Actual deception is often involved in this lying in wait, for the cat appears to be looking in an entirely different direction while she creeps up noiselessly with snakelike movements Then comes the spring on the object, which is clutched with the teeth from above and the claws at the sides. If the object is quite near, or if it hangs like a suspended string, grabbing at it with the claws is substituted for this process.

We may safely assume that the cat does not recognise mock prey as such at this early stage, but we can not be sure, on the contrary, that she thinks it is real prey.

The sight of the moving object is sufficient to account for the whole series of instinctive acts, without calling into use any higher psychological accompaniments. I am therefore not far from right if I use the play with its own or the mother's tail as an illustration here for the sake of simplicity. So far as it is not experimenting, it belongs in the category of chase plays. I have slightly abridged Brehm's beautiful description: "The playfulness of cats is noticeable in their first infancy, and the mother does everything in her power to encourage it. She becomes a child with her children from love of them, just as a human mother forgets her cares in play with her darlings The cat sits surrounded by her little ones and slowly moves her

organe, vol vii. 1894, p. 326. *Cf.* Schneider, Vierteljahr. für Wiss. Philos., ii, 1878.

tail, which Gesner regards as the indicator of her moods. The kittens hardly yet grasp its language, but they are excited by the motion, their eyes take on expression, and they prick up their ears. One and another clutch awkwardly after the moving tail, one tries to clamber on its mother's back and turns a somersault, another has spied the movement of its mother's ears and busies itself with them, while the fifth goes on placidly sucking. The contented mother quietly submits to it all."

I believe that all higher psychological accompaniments are wanting in the first play of young animals, such as with a block or ball or anything of the sort, but are necessarily developed by constant repetition of the game. If a cat keeps running after such a ball, in time a sort of rôle-consciousness comes to her, something like that which accompanies human actions that are intentionally make-believe. This "doing as if," or playing a part, will appear very important in our later observations, and I think we may be sure that the kitten possesses it in some degree at least after frequently repeated experiments. A circumstance that I have not yet mentioned seems to increase this probability. When the ball stops rolling the kitten starts it up again by a gentle tap with her paw, in order to begin the game again. This seems like conscious self-deception, involving some of the most subtle psychological elements of the pleasure that play gives.

Dogs, too, are inclined to chase any moving object. Brehm includes it among their characteristics. "They all run after whatever goes quickly by them, be it man, passing wagon, ball, stone, or what not, attempt to seize and hold it, even when they know perfectly well that it is a thing of no use to them." Every one knows the ridiculous way in which a young dog will chase his own tail, faster and faster, until he falls down. A suspended cord is a welcome plaything to him, too; if he finds he can not pull it down, he seizes it in his teeth and jerks it from side to side, with threatening growls. Close observation of such actions clearly reveals their instinctive origin. The way young dogs will shake a cord or scrap of cloth is excellent practice for shaking their prey—a habit which apparently has the double object of stunning the victim and deepening the hold of the dog's teeth.

The fact that dogs beg to have a stone, a piece of wood, or a ball thrown for them, shows how greatly their chasing impulse is excited by the sight of moving objects. While his master is getting ready for the throw the dog stands waiting with eager eyes and all ready for the spring, and as soon as the object flies off he is after it and trying to seize it. Small dogs seem to hold their prey entirely with their teeth, while my St. Bernard leaps upon the object with his fore paws stiffly extended and deals a blow which would break the backbone of small animals. He will gnaw for a long time on a piece of wood that he has run after, carrying it away in his mouth as he would real prey, and clinging to it energetically if any effort is made to get it from him. This instinct makes it easy to train dogs to carry sticks or baskets. We can be much more confident that a dog has some consciousness of the pretence of the thing in his play than we are in the case of the cat. He knows perfectly well that the stick which he brings and lays at his master's feet time and time again is not alive, and he, too, sets his plaything in motion when there is no one to throw it for him, by seizing it in his mouth and tossing it up in the air. Many dogs delight to play with the feet of their master or mistress

—a black boot has a particular fascination for the rat terrier. It is a pretty sight to see one of them push back a lady's skirt with his paw to find her foot and then pounce upon it eagerly, never biting to hurt her, however—another proof of the consciousness of make-believe. Examples of such play with lifeless objects are not abundant in the literature of the subject that I am acquainted with. However, I am able to cite a few.

Monkeys like to play with balls and other moving objects,* and, according to Rengger, young jaguars do the same, and often play for hours at a time with bits of paper, oranges, or wooden balls.† Captive bears, too, play with blocks and balls. Brehm says that young ocelots "taken young and with care, are very tame; they romp together like kittens, playing with a bit of paper, a small orange, and such objects"; and Hudson says of the puma that at heart it is always a kitten, taking unmeasured delight in its frolics, and when, as often happens, one lives alone in the desert, it will amuse itself by the hour fighting mock battles or playing at hide and seek with imaginary companions, lying in wait and putting all its wonderful strategy in practice to capture a passing butterfly. A tame puma that Hudson knew was delighted when a string or handkerchief was waved before him, and when one person was tired playing with him he was ready for a game with the next comer.‡ Many observers tell us of cranes: these remarkable and intelligent birds throw stones and bits

* Scheitlin, Thierseelenkunde, ii, p. 125. Darwin, Descent of Man. *in loc.*

† Säugethiere von Paraguay, pp. 173, 200, 211.

‡ The Naturalist in La Plata, p. 40.

of mud in the air, as dogs do, and try to catch them as they fall.*

4. *Fighting Plays.*

Such plays are usually to be regarded, in my opinion, as preparatory for the struggle for the female, though there are other reasons for the teasing and tussling of young animals. That pleasure in possessing power that appears in experimentation is certainly present here as well, and such fights serve also as practice for later battles other than those of courtship. Most animals, and especially carnivorous ones, are as pugnacious in conducting their games together as they are over actual prey, for their chasing games very easily lead to fights. But when we reflect that the defenceless creatures, whose only safety is in flight, fight among themselves just as much as the beasts of prey do, we seem to be shut up to the view that the principal use for playful contests is preparation for the later struggle for the female. The close connection between cruelty and pugnacity on the one hand, and sexual excitement on the other, is a fact confirmatory of this view. It is well known that there is a kind of voluptuous pleasure in cruelty. Preyer has published cases of perverted sexual feeling† where the highest degree of excitement was expressed by cruelty to smaller animals; and among some animals—hares, for instance—it is common for the female to be seriously abused in the act of pairing. Schaeffer says:‡ "Fighting and the impulse to kill are

* Scheitlin. Thierseelenkunde. i, 74 Naumann, Naturgeschichte der Vogel Deutschlands. ix. pp. 362, 393.

† Munchen. med. Wochenschrift. 1890. No 23.

‡ Zeitschrift für Psychol. und Physiol. d. Sinnesorgane, vol. ii (1891), p. 128.

so universally attributes of the male animal that we can not doubt the connection between this side of the masculine nature and the sexual. The writer himself believes, from personal observation, that in the perfect male the first shadowy unrecognised suggestion of sexual excitement may be aroused by reading of hunting and fighting, and that the necessity for some sort of satisfaction gives rise to combative games, such as the ring fights of boys." If Schaeffer means that "the fundamental sexual impulse for the utmost extensive and intensive contact of the participants with a more or less clearly defined idea of conquest underlying it" is the main thing, I can only partly agree with him. My idea is that teasing and fighting are closely connected with the sexual life from the fact that they furnish practice for the contest of courtship, without being in any sense satisfying to the sexual instinct. Among many animals that play in this way the female yields to the victor of the males without resistance; and, besides, it frequently happens in the fighting of birds that there is no direct contact at all. Then, again, many young animals have special plays connected with pairing besides their fighting plays.

(*a*) Teasing arises when the desire to fight either does not seek or can not find direct satisfaction. A belligerent animal delights to provoke others that are perhaps not thinking of fighting. After establishing its supremacy by this means the teasing is apt to develop into cruel torture. There are some boys who can not resist dealing an unprovoked cuff to another boy, or pulling his hair, and there are just such animals When Bennett tried to bring an ape to Europe there were other monkeys on the ship that would have nothing to do with him, and he took revenge by seizing them by the tails and dragging them about. He carried one poor fellow to the top of a mast in this way and let him fall. Brehm describes the behaviour of baboons toward two Java apes. "These baboons, like all of their kind, were most jovial fellows, and took the greatest delight in teasing and tormenting the apes, which crouched close together, clinging to one another. The baboons flew at them, tore them apart, poked them in the ribs, pulled their tails, and tried in every way to break up their devoted friendship. They climbed over them, tugged at their hair, forced themselves between the inoffensive pair, until the frightened creatures sought refuge in another corner, only to be followed by their tormentors and maltreated afresh."

A female of the same kind that Brehm brought to Germany loved to tease the snappish house dog. When he took his midday meal out in the court and had stretched himself as usual on the greensward, the roguish monkey would appear, and, seeing with satisfaction that he was fast asleep, seize him softly by the tail and wake him by a sudden jerk of that member. The enraged dog would fly at his tormentor, barking and growling, while the monkey took a defensive position, striking repeatedly on the ground with her large hand and awaiting the enemy's attack The dog could never reach her, though to his unbounded rage, for, as he made a rush for her, she sprang at one bound far over his head, and the next moment had him again by the tail.

"A raccoon that was kept on a farm with other tame animals," writes L. Beckmann, "was specially attached to a badger which was in the same inclosure. On hot days the latter was accustomed to take his nap in the open air under the shade of an alder. Then the mischievous 'coon found his opportunity, but as he feared the badger's bite he carefully kept his distance, satisfying himself with touching his victim softly in the rear at regular intervals. This was enough to keep the sleepy badger awake and reduce him to despair In vain he snapped at his tormentor; the wary 'coon trotted to the edge of the inclosure, and scarcely had the badger composed himself again before he was at his old tricks." I know from experience that young horses often tease their masters. They will run up, stand very quiet with head held high, then spring back and turn with a menacing air. Scheitlin thus describes their actions: "A young horse chased a company of travellers in a narrow Alpine valley. He allowed them to walk past him undisturbed at first, then galloped after them, suddenly stood still threateningly, then turned back and pretended to graze, but soon came bounding on again. This was repeated several times to the no small alarm of the travellers, but he was evidently acting from pure mischief, just like a youth in high spirits." *

Herds of gnus behave in much the same way, so that travellers often have really to run the gauntlet among them.

Saville Kent contributes the following anecdote about dolphins: "A few dog-fish (*Acanthias* and *Mustelus*) three or four feet long now fell victims to their tyranny, the porpoises seizing them by their tails, and swimming off with and shaking them in a manner scarcely conducive to their comfort or dignified appearance. . . . On one occasion I witnessed the two cetacea acting evidently in concert against one of these unwieldy fish (skates), the latter swimming close to the

* Thierseelenkunde, ii, p. 242.

top of the water and seeking momentary respite from its relentless enemies by lifting its unfortunate caudal appendage high above its surface—the peculiar tail of the skate being the object of sport to the porpoises, which seized it in their mouths as a convenient handle whereby to pull the animal about and worry it incessantly."*

Birds, too, give vent to the fighting impulse by teasing one another. Linden reports a parrot that teased others in a good-natured way, and Humboldt had a toucan which delighted in plaguing a sulky monkey that was very easily provoked. Brehm tells this of the ibis: "Those that I have known lived in comparative peace with all the birds that share their quarters, but assumed a certain authority over the weaker ones and seemed to take pleasure in teasing them. The flamingoes especially they could not let alone, and took the strangest way to torment them. As they were sleeping with head buried in their feathers, the ibis softly stole up and picked at their web feet, with no intention of hurting them, but from pure mischief. When a flamingo felt this annoying tickling he moved off, gave a startled glance at the ibis, and tried to get another nap, but his tormentor was soon after him and at the old game."

(*b*) Tussling among young animals. Before entering fully on this part of my subject I am going to cite a case that is to some degree problematical, to prove that I do not overlook the possibility that fighting play may be entirely due to the preying instinct of a certain class of animals I refer to the mock battles of ants. Büchner writes: "It is on the gymnastic exercises and plays of the *Pratensis* that Huber founded his

* Nature, vol. viii, Intellect of Porpoises.

celebrated observations. He saw these ants collect on bright days on top of their hills and behave in a way that he could only describe as regular ring games. They rose on their hind legs, seized each other with fore-feet, feelers, and jaws, and actually wrestled, all in quite friendly fashion. When one gained the ascendency she would seize all the rest, one by one, and throw them over in a pile like skittles Then she dragged them about in her jaws." This description of Huber's was published in many popular papers, but won little credence from the reading public. "Indeed, I myself," says Forel, "found it hard to believe, in spite of the accuracy with which Huber recorded his observations, until I saw it myself." A colony of *Pratensis* gave him this opportunity as he approached them carefully. The wrestlers seized one another with feet and jaws, rolled together to the ground, just as playful urchins like to do, pulled each other into their holes only to come out and begin over again. All this was apparently done without anger or spiteful feeling; it was clear that they were actuated only by a spirit of friendly rivalry.* Supposing that this is all play,† an admission that I am not altogether prepared to make, there is, of course, no connection with courtship. "I can understand," says Forel, "that it must appear all the more incredible to those who have not seen it, when they reflect that sexual instinct can have nothing to do with this play." The mock fights of ants must then be entirely for practice preparatory to their unusually quarrelsome and predatory way of living.‡ Notwithstanding, I must hold

* A. Forel. Les fourmis de la Suisse, 1874.

† Büchner, Aus dem Geistesleben der Thiere, pp. 196, 220.

‡ McCook and Bates also have observations on the play of ants. See Romanes, Animal Intelligence, p. 88.

to the belief that mock fighting in general is preparatory for the courtship contest. The fact that ants form an exception does not warrant the conclusion that the principle does not apply to the animals referred to in what follows.

Again, I begin with the dog. All kinds of puppies are indefatigable in playful romping, and gain in this way much that is needful in the serious struggles of later life. While they are very young, little dogs chase each other awkwardly and try to seize the throat. Fox terriers usually try to dodge the first attack,* others rise on their hind feet and fight with front paws and teeth. When one is thrown he at once turns on his back to protect his neck, and dexterously wards off the enemy with his fore feet. The victor, equally skilful, stands with feet wide apart over his fallen foe and prevents him from rising. If the dogs are of unequal size, the big one often lies down of his own accord and carelessly keeps the little one at bay, as he makes excited dashes for the enemy's throat from all sides. The quiet movements of a huge mastiff in contrast with the audacity and violence of a terrier, which attacked him in this way, have often amused me.

Tussling like this, where pleasure in the possession of power and the closely related rivalry, as well as mere pugnacity, play important parts, is almost universally practised among animals. All the feline tribe without exception indulge in it, young tomcats especially, so that the Germans have a special word for their fighting, "*Katzbalgerei.*" At the age of two months young lions begin their play, which is like that of the house cat, and the same is true of tigers, jaguars, leopards,

* *Cf.* Diezels, Niederjagd, p. 506.

ocelots, etc. Young wolves howl and yelp during their play; when tame they play with children. Brehm writes: "Hyenas, taken young, soon become accustomed to a particular person, and have a method of showing their pleasure at the appearance of a friend, that is not employed by any other beast of prey that is known to me. They rise with cries and jump about like mad, struggle with each other merely from pleasant excitement, bite one another, roll over and over on the ground, spring and leap and hop about the cage, all the time keeping up uninterruptedly a sound for which there is no word—the nearest approach, perhaps, is to call it a twittering." Young male weasels romp and tussle, sometimes biting one another severely, when the savage nature asserts itself. Sables often play merrily together, standing upright the better to fight, and I have seen two ant-eaters chasing and plaguing each other. Bennett says of young duckbills: "One evening my two little pets came out as usual toward dusk and ate their supper. Then they began to play like a pair of puppies, seizing one another with their bills, striking with the fore paws, clambering over each other, etc. When one of them fell in the strife and the other confidently expected him to get up at once and renew the battle, if it occurred to him to lie still and scratch himself, his comrade calmly watched the proceeding and waited till the play began again."

Bears stand upright when they fight, like squabbling boys. A young polar bear that I have watched was fond of playing with his mother; he chased her, bit her feet, and scratched her nose, while she tried to seize him as he lay on his back. Badgers "come out on still, sunny days and amuse themselves; the clumsy young ones hug each other like bears, tussle, and roll about, dealing cuffs right and left." *

Beckmann describes the actions of a tame young raccoon so beautifully that I can not resist quoting the whole passage: "He had formed an offensive and defensive alliance with a large bird dog. He was quite willing to be tied to it, and then both followed their master step by step, though when the raccoon was alone on the chain he constantly pulled away. As soon as he was unchained in the morning he joyfully bounded off to find his friend. Standing on his hind feet, he threw his fore paws around the neck of the dog, whose head he gently bent forward. Then he examined and sniffed about his friend's body with curiosity and interest, seeming to discover new charms daily. Where the hair was rough, he carefully licked it down. The dog stood motionless and strangely serious during the whole inspection, which frequently lasted a quarter of an hour, changing his position or raising a limb when the raccoon indicated that it was necessary. He drew the line, however, on having the creature mount on his back, and the attempt was a signal for a prolonged tussle, where much courage and dexterity were displayed. The raccoon's mode of attack was to spring in an unguarded moment at the throat of his much larger and stronger opponent. Thrusting his body between the dog's front legs, it attempted to hang on by his neck. If he succeeded in this the dog was worsted and could only roll frantically on the ground in his endeavours to rid himself of the fervid embrace. To the credit of the rogue it should be said that he never

* A. and K. Müller, Wohnungen, Leben und Eigenthümlichkeiten in der höheren Thierwelt, p. 62.

abused his advantage, but contented himself with keeping his head close under the dog's throat, out of danger of a bite."

I have already referred to the fact that animals not inclined for fighting, except for defence, are as fond of playful contests in their youth as are the most dangerous and aggressive beasts of prey. In such cases we must expect to find in preparation for courtship the leading if not the only reason for such fighting. Young horses, donkeys, zebras, etc, tear madly over the plains, rear up at each other, strike with head and fore feet at one another's legs and neck. Calves, too, fight obstinately, approaching each other with lowered head, each trying to push the other back. Goats fight in the same way, and they too often measure strength in friendly rivalry. If the contest becomes earnest, they commonly rise on their hind feet and exert all their strength for a side push.

I have seen two Madagascar monkeys wrestling together just as dogs do, except that the play became more complicated from their being able to hold on with hands and feet.

Every one knows how lambs frisk and play about a meadow. Kids play just as the goats do, while young deer rise on their hind feet and strike out with the front ones.*

According to Steller, young sea bears also play and quarrel like puppies. The father stays by and watches them, and if a quarrel begins in earnest he urges them on with growls, and kisses and licks the victor, then pushes him to the ground, and is pleased if he resists. It is worthy of remark that seals, whose young, it seems,

* A. and K. Müller, Thiere der Heimath, i, p. 422.

universally indulge in vigorous mock contests, are especially passionate and pugnacious during their courtship.

Finally, we will take a few examples from the birds. Water-wagtails chase and bite each other, apparently in play, as is seen "most commonly late in summer among young birds." * Young house and field sparrows peck one another soundly while they are carrying on their courtship plays, as do the nuthatch, starling, wood lark, water-wagtail, and goldfinch. Young partridges stands with wings wide spread and fight as hotly as if they were already contesting for a lady love.†

(*c*) Playful fighting between adult animals. Many a grown animal still takes pleasure in the mock combats that he learned in youth, and it is unnecessary to dilate on the usefulness of such sportive measuring of strength in keeping him fit for actual warfare. From a psychological point of view, however, this phenomenon is especially noteworthy from the fact that the adult animal, though already well acquainted with real fighting, still knows how to keep within the bounds of play, and must therefore be consciously playing a rôle, making believe. This can hardly be denied, I think, in some of the following cases.

Finsch says that seals make so much commotion in the water while playing that they appear to be fighting angrily, 'though it is really all frolic, just as the biting is in which they indulge on land. Two of them open their powerful jaws, angrily howling in a fearful way, as though a serious combat were about to take place, but instead they lie down peacefully side by side, and perhaps begin mutual lickings."

---

* Naumann, iii, p. 814.

† Chr. L. Brehm, Beiträge zur Vögelkunde, ii, p. 748.

Friendly dogs often keep up their playful fights to an old age without ever being in the least angry; and among the cattle on Alpine pastures, where the greatest freedom is allowed them, these playful contests are frequent. "The Alpine cows," says Scheitlin, "learn to know their proper food more quickly, are more good-natured, and take more pleasure in life than others. They fight valiantly, both in play and in earnest; with all their amiability and fondness for one another, they gore and push terribly, yet not in anger or bad temper, but like a lot of boys that fight to exercise their muscles. They will stand for a long time with heads lowered and horns interlocked, as if they would never separate They do not look one another in the eye, as men do, when fighting; their eyes are on the ground, their whole mind is concentrated on the push When one succeeds in shoving the other back, neither seems to care; the loser is not in the least ashamed, nor does the victor show any pride or pleasure. Some of them are very pugnacious, and display great courage and persistence." *

Females are thus seen to display the eagerness for combat that is in general so much more the characteristic of the male; just as among ourselves, masculine instincts often appear in women. Some female cats are twice as aggressive and bloodthirsty in their breeding time as any male, and there are some kinds of birds whose females imitate the song of the males and mingle in their battles.†

* Scheitlin, Thierseelenkunde, ii, p. 201.

† The cows that undertake to lead and rule the herd will sometimes fight to the death. Their leadership is like that of the bull in a wild state. See Tschudi, Das Thierleben der Alpenwelt, 1890, p. 542.

Pechuël-Loesche tells us, in the report of the Loango Expedition, that African sheep are much more courageous and bellicose than the European varieties. The ram Mfuka that the travellers kept at their station seems to have been a regular tyrant. "He would not endure quarrelling or noise among the men or animals. When the amorous goats fought, he would look at them inquiringly for a while and then deliberately run them down. If the men quarrelled, he acted as peacemaker in the same thoroughgoing way, much to the amusement of all concerned. On one occasion the spokesman of an inland chief was talking violently before the door, when Mfuka gently came up, measured his distance, and dealt a mighty blow so energetically on the solidest part of the man's anatomy that he fell sprawling on the sand. That put an end to the speech. It was a rare spectacle to see the amazed ambassador sitting there, and the ram standing by solemnly gazing at him." *

Brehm says of two curly bears, a male and a female: "Soon began the merry game, in which they whirled about so that it was impossible to distinguish one from the other. They rolled on the floor like balls, seizing and hugging each other, using jaws and tails indiscriminately as weapons of offence and defence." It is noteworthy that they never paired, though Brehm hoped they would, and their play seems therefore to have only the significance that Schaeffer attaches to such romping

Now a few examples from the birds. The hooded raven, which Naumann watched from his hiding place for hours at a time, is a very lively bird "They often quarrelled, but never seriously; they danced and hopped,

* Loango Expedition, iii, 1, p. 301.

rolled over in the snow, lay on their backs, took constrained positions, and uttered strange cries, apparently with great effort." *

Sale, who brought the first kakapo to Europe in 1870, writes of this bird: "His sportiveness is remarkable. He runs from his corner, seizes my hand with claws and beak, and tumbles about like a kitten on the floor, still holding the hand, then he hurries off as if to prepare for another attack. He is sometimes inclined to be a little too rough in his play, but a mild reproof checks him, and he is really an amusing fellow. When I tried the experiment of bringing a dog or a cat to his cage he would dance up and down with wings outspread and making every pretence of anger, and his pleasure was evident when he succeeded in exciting the animal." To me it is very doubtful whether this was in truth only feigned anger. Naumann also regards the following familiar phenomenon as a play: "It is fine to see how the jackdaws amuse themselves during a strong wind at the top of a tower or tall tree. One will hustle another off and take his place, only to be pushed off in his turn by the next comer, and so on for hours. Crows often do this too." †

Perhaps Brehm's report of a buzzard in captivity belongs here also. This bird made friends with a little dog, perching between his feet when he lay down, frolicked with him, and tweaked his hair with its beak. Baldenstein had a tame vulture that was very fond of him. Even when he teased the bird it made only playful attacks on him, though under other circumstances it made terrible use of its dangerous weapons.

The question now arises whether such playful fight-

* Naumann, ii, p. 69. † Ibid., ii, p. 96.

ing as we have been considering ever occurs during the breeding season. The contest for the possession of a female is usually a serious matter, often a life-and-death struggle, and yet may there not be some fighting connected with this period that is playful? Here, as in most questions of animal psychology, absolute certainty is unattainable, but we may inquire into the probabilities, and it seems to me not impossible that contests playful in character may take place even during courtship. Perhaps I may be allowed a human instance. The belligerent spirit of young peasants is certainly of this nature, little as the brawlers are conscious of it. And however serious the fights that arise on Sundays and holidays, they impress us as at bottom playful, for neither combatant wishes actually to injure the other, but rather to prove his own superiority, though this may involve a desperate struggle. The fencing of students, too, although often resulting in injuries that would be dangerous without the immediate service of a surgeon, are yet avowedly for sport. It occasionally happens that a desire for revenge leads to the inflicting of intentional and serious injury, but as a general thing it is all for practice in acquiring skill and courage for use in more serious circumstances. It may be the same with animals. Even when they have overstepped the bounds of the friendly tussling that we have been considering, and the contestants are angry and really trying to hurt one another, still there may be something of the temper of play. I do not assert that this is often the case, but I may give a couple of examples that at least give colour to the idea. We often see grown dogs chase each other with loud cries without coming to a fight at all, and this before the very eyes of the object of their rivalry. While snappish dogs bite one another sharply, it seems to be done chiefly to prove how formidable they are and how fearless. They slowly come together with stiffened legs, back up, and ears and tail erect; and each seeks to determine by characteristic and comic sniffing what sort of fellow he has to do with. Then they slowly walk around each other for some time, keeping the legs stiff and each with his head turned, as if aiming an attack at the other's throat. Even after all this they are very likely to separate quietly, but sometimes they come to open combat. With frightful screams they leap at one another, show their teeth, growling, and sometimes bite a little, but almost always part without having gone to the length of a serious struggle.

My other example is from Baldamus's description of the night heron: "When no marauder disturbs them they find means to torment one another, chasing and fighting with loud cries. They have a peculiar game of climbing, during which they sometimes get into the most ridiculous situations and scream constantly. For example, while a female is busy appropriating a twig or some such matter from a neighbouring nest, it occurs to the male to pick at the feet of a bird standing above him The offended one spreads his wings threateningly, opens his beak, and tries to retaliate, but is so closely pressed by the aggressor that he retreats until the end of the limb is reached or the courage of despair inspires the victim. The amusing feature of it lies in the contrast between the extravagantly threatening aspect of the aggrieved bird and his trifling efforts at defence. The wide-open beak, the constantly varying cries, 'Koau! krau! krau!' etc., the flaming eyes, red and flashing with rage, the wings raised so threateningly, the head alternately drawn back and protruded, the extraordinary contortions of the whole body, the erection of the head and neck feathers—all this leads one to expect a life-and-death struggle, and behold! they scarcely do more than touch each other with the tips of their wings, very rarely with the beak. They rage and storm like Homeric gods, but with no result."

According to Darwin, a competent observer goes so far as to say of the *Tetrao umbellus:* "The contest of the males was only a pretence arranged to display themselves advantageously before the admiring females collected near, for I have never been able to discover a mutilated hero, and seldom one with more than a feather turned."* Brehm and Naumann† both contribute to the following description of the remarkable behaviour of the willow wren, sometimes called the fighting wren, which before the pairing is a particularly peaceable bird: "But this quality disappears entirely as soon as the pairing time arrives; it is now that they deserve their second name, for the males fight continually and with no apparent cause ‡—if not over the female, over a fly, a worm, a beetle, a place to perch, anything or nothing. It is just the same whether females are present or not, whether they enjoy absolute freedom or are in captivity, whether they have been taken a few hours ago or have lived in a cage for years. In short, they fight at all times and under all circumstances. When free, they collect at an appointed spot; usually a moist elevation covered with short grass and about two metres in diameter is chosen for the arena, and is resorted to several times daily by a certain num-

* Descent of Man, ii, p. 48.

† Naumann, vii, p. 535.

‡ Among students, to be called a dunce, to be jostled, or even gazed at, is cause enough.

ber of males. . . . The first arrival looks anxiously about for a second, but when he comes, should he prove not exactly fit, a third and fourth are awaited, and then the battle opens. Each having found his antagonist, they fall to, fly at each other, and fight vigorously till they are tired, when each returns to his place to rest and collect his strength for the next round. This goes on till they are exhausted and retire from the field, to return soon, however, in most cases More than two never fight together, but if a good many are on the ground at once, as often happens, they fight in pairs, and cross one another in such marvellous leaps and bounds that a spectator at a little distance would think the birds were possessed of an evil spirit, or else gone crazy.

"When two of these birds come upon a grain at the same time, they both stand still at first, trembling with rage, then stooping so that the hind part of the body is higher than the head, and ruffling up their feathers, they dart at each other, dealing sharp blows. . . . Sometimes a female comes to the battle ground and takes a place with the fighting males, yet she does not long mix in the strife, but soon goes away. It may happen that a male accompanies her and stays with her, but two males never leave together or chase one another on the wing. The battle is fought out on the ground, and then peace is established."

5. *Constructive Arts*

It is true that very few of the phenomena connected with building by animals have anything to do with the psychology of play, but before taking any further steps it is necessary to inquire into the part played by instinct in the exercise of constructive skill by the higher animals, and especially by birds.

Wallace, in his Philosophy of Birds' Nests, has tried to prove that inherited instinct has very little to do with it. The material, he says, depends on circumstances, and the form partly on natural impulse, but chiefly on imitation. The young bird lives in the nest for days and weeks and learns to know its every detail. During the time he is learning to fly he studies the outside, and naturally keeps a memory picture of the parental home against his own time of need, when he imitates it. The manner of building which has become *tradition* through imitation, among savage tribes, is thus seen also among the higher animals.

Worthy of respectful consideration as these opinions undoubtedly are, it is extremely probable, to say the least, that Wallace has gone too far. Though here and there imitation may play a more or less important part in this work, it would be hard to dispense with the idea that hereditary impulse is, as a rule, responsible for the constructive skill of animals. The making of a chrysalis by the moth is so unquestionably instinctive that no one will deny it, and such facts among the lower orders naturally lead us to consider the case of higher animals analogous. It should be borne in mind, too, that young birds of the kinds that nest but once can not in this way learn the manner of constructing a nest, since the finished one shows little of the process. Weir wrote to Darwin in 1868: "The more I reflect on Mr. Wallace's theory that birds learn to make their nests because they have been themselves reared in one, the less inclined do I feel to agree with him. . . . It is usual with many canary fanciers to take out the nest constructed by the parent birds and to put a felt nest in its place, and when the young are hatched and old enough to be handled, to place a second clean nest, also of felt, in the box, removing the other, and this is done to avoid *acari*. But I never knew that canaries so reared failed to make a nest when the breeding time arrived. I have, on the other hand, marvelled to see how like a wild bird's their nests were constructed." *

It is, of course, difficult to determine how much is due to instinct and how much to intelligence, for no one claims that the building of higher animals is *purely* instinctive. Take, for example, Naumann's beautiful description of the skilfully made nests of the golden oriole, so like an inverted nightcap: "One of them (usually the male) comes flying with a long thread or grass blade in his bill and tries to fasten the end of it to a bough, perhaps with the help of his spittle, while the female catches the loose end and flies with it two or three times around the bough and fastens it in the same way to a forked limb opposite." †

This can not be all instinct; it is a case where inherited instinct and individual experience work together. The Mullers have expressed their belief that, though old birds usually build better than their young when there is any difference at all, still the instinct for building is, after all, a gift of Nature.‡

"The ravenous screeching young owls do not think of making studies in architecture. . . . If the parents have a second brood, the young of the first never come near them, nor does it enter their heads to take lessons in building.#

* Romanes, Mental Evolution in Animals, p. 226.

† Naturgeschichte der Vögel Deutschlands, ii, p. 181.

‡ A. and K. Müller, Thiere der Heimath, i, p. 39.

Ibid., i, p. 125.

"As a matter of fact, no naturalist has yet been able to prove that old birds instruct their young in nest-building. It would be impossible for those that nest but once, as the young can not be present when the parents build; yet the next spring, when they are only a year old, they go about the construction of their own nest with as much assurance as if they had been in the business a long time." *

I cite Naumann next, who plainly indicates the two-fold nature of the phenomenon: "We may well wonder at the mysterious instinct that enables young birds to build at their first attempt nests as perfect as those of their parents, and similar to them in material, position, and form; but it can not be denied that their art can be brought to even greater perfection by means of practice." †

In inquiring now as to the connection between these arts and the psychology of play, it becomes apparent that building in general is not playful. The earthworks of beavers, foxes, badgers, fish-otters, rabbits, etc., the leafy arbours of many kinds of apes, the nests of the perch, hedgehog, squirrel, field mouse, and bird serve a purpose that is directly useful. But since all art has at least some likeness to play, it follows that building of this kind is not properly called art, any more than the rude shelters of our primeval ancestors can be called products of architecture. Only in special cases, then, can we speak of playful building Darwin sees such a case in the well-known fact that caged birds often build nests for amusement, when they have no occasion to use

* A. and K. Müller, Wohnungen, Leben und Eigenthümlichkeiten der höheren Thierwelt, 1869, p. 216

† Naturgeschichte der Vogel Deutschlands, i, p. 97.

them.* The weaver bird offers the most familiar example. Carus, too, speaks of the plaiting "which many birds work at if prevented from building nests for themselves. It is especially interesting to watch the *Ploceus sanguinirostris,* now so common in Europe, when it can not build its peculiar purse-shaped nest, how it makes use of every available scrap of thread or straw in interweaving and adorning the bars of its cage. Surely this bird evinces a certain intelligence, which is not of the lowest order, as any one must be convinced who watches it at work for any length of time—how it holds a thread in its claw, seizes it with the beak, pushes it through the grating, ties a good knot, and proceeds to weave it in and out." †

This might be regarded as a kind of play, depending, however, upon the abnormal conditions of the bird's life. But for its artificial *milieu* it would build a nest, and since instinct forces it to build something, its activity assumes a playful character, owing to circumstances imposed by man. The attempts of some male birds to build nests on their own account, before they have assumed the responsibility of wedlock, may, however, be regarded as purely playful. The Müllers tell us that the wren does this, sometimes making two or three nests imperfectly alone, before he unites with the female in building the one on which she sits. "This haste to build," says the observer, "is nothing but happy sportiveness on the part of the little creature bewitched by love." It is probably due to the fact that the awakening of sexual passion arouses all the instincts connected with it to activity. Many birds pick on the ground during their

* Descent of Man, ii, p. 52 f.

† C. G. Carus, Vergleichende Psychologie, 1866, p. 213.

courtship, as if trying to take something up, others will throw little stones behind them, and still others carry about on their beaks a small feather of the adored one. The action of the wren described above is only one step further in the same direction, and we find its culmination in the wonderful pleasure-house of the bower bird. Another manifestation of it is found in the fact that during the time of their courtship many female birds allow themselves to be fed by the male, just as the young are later on.

But more important for our purpose are the strange methods of building ornamentation employed by some animals. If no other meaning can be discovered for them, they may very properly be regarded as playful. I know of only two instances in mammals, and the first of these is imperfectly vouched for and dubious. Darwin says that the viscacha, a South American rodent, has the remarkable habit of collecting at the mouth of its burrow every portable object within its reach, so that heaps of stones, bones, thistle stalks, lumps of earth. dry dung, etc., are found near their holes. It is even related of a traveller who lost his watch in the region that he recovered it by searching among the viscacha mounds along the way.*

Hudson corroborates these reports, and finds a use for the habit: "For as the viscachas are continually deepening and widening their burrows, the earth thrown out soon covers these materials, and so assists in raising the mounds," which protect their dwellings from overflow.† He further remarks that these animals always build in an open plain, on even, close-shaven turf, where

* Darwin, Journey around the World.

† The Naturalist in La Plata, p. 304.

an approaching enemy can easily be descried. This instinct for clearing the ground Hudson considers sufficient to explain the collection of objects lying about. If Hudson is right, as seems probable, there is, of course, nothing playful about it. Darwin, on the contrary, thought this habit of the viscachas analogous to that of certain birds which I will now describe. The Australian atlas bird (*Calodera maculata*) builds an intricately woven structure of twigs to play in, and collects near it shells, bones, and feathers, especially those brightly coloured. Mr. Gould says that when the natives lose any small, hard objects they at once search these places, and he knew of a pipe that was recovered in this way.*

If Darwin regarded these as the only examples of the kind, he must have overlooked some familiar instances. One at least relating to mammals is cited by James from Lindsay's Mind in Lower Animals. Referring to a nest of the Californian wood rat, which he discovered in an unoccupied house: "I found the outside to be composed entirely of spikes, all laid with symmetry, so as to present the points of the nails outward. In the centre of this mass was the nest, composed of finely divided fibres of hemp packing. Interlaced with the spikes were the following: About two dozen knives, forks, and spoons; all the butcher's knives —three in number—a large carving knife, fork, and steel; several large plugs of tobacco, . . . an old purse, containing some silver, matches, and tobacco; nearly all the small tools from the tool closets, with several large augers, . . . all of which must have been transported some distance, as they were originally stored in distant parts of the house. . . . The outside casing of

* Darwin, *loc. cit.*

a silver watch was disposed of in one part of the pile, the glass of the same watch in another, and the works in still another." *

The other examples are of birds. The so-called thieving of crows and ravens shows their characteristic bent in its simplest form, for they all delight to carry small, bright objects to their nests. Naumann certifies to it of the pond raven, crow, hooded raven, curlew, jackdaw, and magpie.† The bastard nightingale also likes to trim the outside of its nest with bark, feathers, shavings, and scraps of paper.‡ The Müllers describe a wren's nest that was lined with bright yellow chicken feathers.# Romanes says that there are "many species of birds that habitually adorn their nests with gaily coloured feathers, wool, cotton, or other gaudy material. . . . In many cases a marked preference is shown for particular objects, as, for instance, in the case of the Syrian nuthatch, which chooses the iridescent wings of insects, or that of the great crested flycatcher, which similarly chooses the cast-off skins of snakes. But no doubt the most remarkable of these cases is that of the Baya bird of Asia, which, after having completed its bottle-shaped and chambered nest, studs it over with small lumps of clay, both inside and out, upon which the cock bird sticks fireflies, apparently for the sole purpose of securing a brilliantly decorative effect. Other birds, such as the hammer-head of Africa, adorn the surroundings of their nests, which are built upon the ground, with shells, bones, pieces of broken glass and earthenware, or any objects of a

* James, The Principles of Psychology, ii, p. 424.

† Naturgeschichte der Vögel Deutschlands, ii, p. 50.

‡ A. and K. Müller, Thiere der Heimath, i, p. 56.

Ibid., i, p. 61.

bright and conspicuous character which they may happen to find." *

Still more remarkable is the case of the bower bird, which does not, indeed, adorn its nest, but builds a playhouse, in the shape of a tunnel on the ground, entirely for the purposes of courtship, and decorates it in every possible way. Both sexes work in its construction, but the male is the director.

So strong is this instinct that it is practised under confinement, and Mr. Strange has described the habits of some satin bower birds which he kept in an aviary in New South Wales: "At times the male will chase the female all over the aviary, then go to the bower, pick up a gay feather or a large leaf, utter a curious kind of note, set all his feathers erect, run round the bower, and become so excited that his eyes appear ready to start from his head; he continues opening first one wing and then the other, uttering a low, whistling note, and, like the domestic cock, seems to be picking up something from the ground, until at last the female goes quietly toward him." Captain Stokes has described the habits and "playhouses" of another species—the great bower bird—which was seen "amusing itself by flying backward and forward, taking a shell alternately from each side, and carrying it in its mouth through the archway. These curious structures, formed solely as halls of assemblage, where both sexes amuse themselves and pay their court, must cost the birds much labour. The bower, for instance, of the fawn-breasted species is nearly four feet in length, eighteen inches in height, and is raised on a thick platform of sticks." †

* Romanes, Darwin and after Darwin, i, p 380.

† Cf. Darwin, Descent of Man, vol. ii, p. 77.

Moreover, these bowers are elaborately decorated, and the manner of decoration differs in the three varieties of birds. "The satin bower bird collects gaily coloured articles, such as the blue tail feathers of parrakeets, bleached bones, and shells, which it sticks between the twigs or arranges at the entrance. Mr. Gould found in one bower a neatly worked stone tomahawk and a slip of blue cotton, evidently procured from a native encampment. These objects are continually rearranged and carried about by the birds while at play. The bower of the spotted bower bird 'is beautifully lined with tall grasses, so disposed that the heads nearly meet, and the decorations are very profuse.' Round stones are used to keep the grass stems in their proper places and to make divergent paths leading to the bower. The stones and shells are often brought from a great distance. The regent bird, as described by Mr. Ramsay, ornamented its short bower with bleached land shells belonging to five or six species and with berries of various colours—blue, red, and black—which give it when fresh a very pretty appearance. Besides these there were several newly picked leaves and young shoots of a pinkish colour, the whole showing a decided taste for the beautiful. Well may Mr. Gould say, 'These highly decorated halls of assembly must be regarded as the most wonderful instances of bird architecture yet discovered,' and the taste, as we see of several species, certainly differs." *

In reviewing these strange practices of birds found in such various parts of the earth, we find that, though here and there an explanation like that of Hudson for the viscachas may be hazarded, in the main no better

* Descent of Man, vol. ii, p. 124.

ground exists for them than the fact that the birds take pleasure in possessing objects that are gaily coloured or bright. Our next question, then, is, Whence arises this delight in bright and gaily coloured things? Since Darwin's time it is the custom to attribute everything of this kind to a direct æsthetic enjoyment of the beautiful. But that is an unsatisfactory explanation, originating in a misconception of the essentials of æsthetics. At the most, such satisfaction as these birds feel can only be regarded as a stimulus to sensuous pleasure, which, strictly speaking, is not æsthetic enjoyment at all For the full perception of beauty, the sensuous pleasure arises first when, through the function which I have called "inner imitation," the sensuously pleasing object takes on spiritual embodiment. It is highly improbable that a psychological operation such as this, which is rarely called forth even in men in its full strength, should be developed in animals. What they really feel is the pleasure of the senses produced by physical well-being without reference to æsthetics, such as may be produced in ourselves by the contemplation of a clear sky, pure air, and green fields. This sensuous delight in what is bright and gay is an important antecedent to æsthetic pleasure because it assures a lively perception of the object, but it should not be mistaken for æsthetic pleasure itself.

Further, we may well suspect that this delight in striking colours and forms is not unconnected with the sexual life. It is well known that Darwin teaches that these characteristics in male birds largely control sexual selection. Later I shall discuss the question whether we can rightly refer the origin of such phenomena to sexual selection, even though its later influence be granted. There can be no doubt that animals are excited in this way by the display of what might be called their wedding finery, but this feeling may very well be extended by association to other and unusual things, all of which the birds are attracted to because of their tendency to produce sexual excitement.* The following anecdote, given to Romanes by a lady, illustrates this:

"A white fantail pigeon lived with his family in a pigeon house in our stable yard. He and his wife had been brought originally from Sussex, and had lived, respected and admired, to see their children of the third generation, when he suddenly became the victim of the infatuation I am about to describe.

"No eccentricity whatever was remarked in his conduct until one day I chanced to pick up somewhere in the garden a ginger-beer bottle of the ordinary brown-stone description. I flung it into the yard, where it fell immediately below the pigeon house. That instant down flew *pater familias*, and to my no small astonishment commenced a series of genuflections, evidently doing homage to the bottle. He strutted round and round it, bowing and scraping and cooing and performing the most ludicrous antics I ever beheld on the part of an enamoured pigeon . . . Nor did he cease these performances until we removed the bottle, and, which proved that this singular aberration of instinct had become a fixed delusion, whenever the bottle was thrown or placed in the yard—no matter whether it lay horizontally or was placed upright—the same ridiculous scene was enacted; at that moment the pigeon came flying down with quite as great alacrity as when

* I find a similar idea advanced in Lloyd Morgan's Animal Life and Intelligence, p. 408.

his peas were thrown out for dinner, to continue his antics as long as the bottle remained there. Sometimes this would go on for hours, the other members of the family treating his movements with the most contemptuous indifference and taking no notice whatever of the bottle. At last it became the regular amusement with which we entertained our visitors, to see this erratic pigeon making love to the interesting object of his affections, and it was an entertainment which never failed, throughout that summer at least. Before next summer came he was no more." *

Romanes agrees with the lady who wrote the description in regarding this as a pathological case, but, even if that is correct, still the actions of this pigeon throw some light on the question we have been considering. In order to estimate their real relation to play we must return to our first division, namely, experimentation. Since seizing, holding, and carrying things about form one manifestation of experimentation, it is natural that an unusual object should excite the attention and give pleasure to animals. A child, too, takes pleasure in collecting bright objects, and the fact that they, as well as some birds—the warbler, for instance—are continually handling their treasures, carrying them from place to place and rearranging them, clearly shows the experimental character of such habits. An instinct very closely connected with experimentation, but not yet mentioned, is involved here, for where we find pleasure in power, pleasure in ownership is not far off. James calls this the instinct of appropriation or acquisitiveness. "The beginnings of acquisitiveness," says he, "are seen in the impulse which very young

* Romanes, Mental Evolution in Animals, p. 173.

children display to snatch at or beg for any object which pleases their attention." * I regard the instinct whose mandate in the struggle for life is, "Keep what you can get," as very important. Men and animals must learn not only to acquire, but also to defend and protect their property with tenacious energy. How purely instinctive this is, is shown by the tame canary that will peck angrily at the hand of even its beloved owner, that has just given it the bit of salad or apple which it now defends.

But there is a playful side to it as well, as witness the stubbornness with which a dog at play will cling to the stick in opposition to his master. As James remarks, the zeal for collecting is the most common form of it among ourselves. "Boys will collect anything that they see another boy collect, from pieces of chalk and peach-pits up to books and photographs. Out of a hundred students whom I questioned, only four or five had never collected anything." † This passion is highly developed among the mentally deranged. Many patients in lunatic asylums have a mania for picking out and treasuring all the pins they can find. Others collect scraps of thread, buttons, rags, etc., and are happy in possessing them.‡ The thieving of jackdaws and magpies is something like this.

Finally, this observation is to be noted. In all the cases we have considered the desire to experiment with or to get possession of objects has been directed to such things as were bright or gaily coloured. Now, if we find in the preference for such things an antecedent

* The Principles of Psychology, ii, p. 422.

† Ibid., ii, p. 423.

‡ Ibid., ii, p. 424. Kleptomania, too, belongs to the pathology of this instinct.

to æsthetic enjoyment, surely the same instinct directed toward building can be regarded as an antecedent to æsthetic production. I find three principles influential in the production of human art: First, *self-exhibition* * (Selbstdarstellung); second, *imitation;* third, *ornamentation.* Now one of these and now another seems to be more important, but it always proves on examination that they are all essential. (I shall have more to say on this point in the last chapter.) The examples I have cited emphasize the principle of ornamentation chiefly, but the other two were present also. The habits of the warbler, for instance, suggest that inherited instinct is not working alone, but is assisted by tradition, for the younger birds seem to imitate what they see their elders do. So it appears that imitation has a part in the formation of any habit where the young prefer, as their model, those of the older ones who have distinguished themselves in the art in question.

Something akin to self-exhibition is discernible too. That feeling which is so plainly shown in the sportive love-making of the bird probably has something to do with the fanciful trimming of his nest. Just as we extend our ego to the ends of our canes and to the top of our high hats, as Lotze says, just as we are vain of well-made clothes, of a fine establishment, of the ornamental façade of our house, or even of the advantages of the neighbourhood in which we live, so the bird may feel a pride in the striking or sensuous pleasing object, that is akin to self-exhibition.

* [An English term suggested by Baldwin (Psycholog. Review, i, Nov., 1894, p. 620), with reference to Marshall's Pain, Pleasure, and Æsthetics, and accepted by Marshall.]

These thoughts are merely thrown out, not as serious statements, nor even as hypotheses, but rather as half-playful speculation as to what may be going on in the bird's mind, and may be taken for what they are worth. However, it remains true that our point of departure, namely, the delight in what is bright or gay, is very remarkable, is a mental capability bringing the animal that possesses it into line with primitive man at this one point, when the development of his other faculties lags far behind. Their case is like that of one of those astonishing, and at the same time stupid, mathematical geniuses, whose mental capacity is inferior to that of the average man in all directions save one, while his ability to grasp and manipulate series of numbers is something phenomenal. But we must bear in mind that such phenomena may after all be explained on grounds of practical utility; and if thus explained they have no place in the psychology of play.

6. *Nursing Plays.*

During the time that I spent in study preparatory to writing this book I naturally became much interested in human play as well, and although my classification of animal play has not to my knowledge been influenced by any system of human play, I confess that I am now confronted by a problem that would not have been likely to attract my attention if I had not seen children at play. We all know how much of that is with dolls, and the question for us now is whether there is anything analogous to it in the animal world. Of course, an animal in its natural condition can never be the possessor of a doll—that is, a plastic representation of an individual of its own kind—and even if one were given him he would not know how to play with it. Romanes relates of the same ape that his sister so admirably described: "I bought at a toyshop a very good imitation of a monkey and brought it into the room with the real monkey, stroking and speaking to it as if it were alive. The monkey evidently mistook the figure for a real animal, manifesting intense curiosity, mixed with much alarm if I made the figure approach him. Even when I placed the figure on the table and left it standing motionless the monkey was afraid to approach it." *

My St. Bernard displayed feelings of curiosity mingled with fear when I held an imitation white poodle before him, and as I made the figure bark his astonishment recalled Schiller's verse on the power of song:

> "Amazed, and with delicious fear
> He heard the minstrel's lay, and hid."

There was no sign of a wish to play with the doll. But this does not dispose of the subject. Real dolls are not the only thing that children play with. Little girls often prefer some make-believe, a comb, a fork, a stone, a bit of bread, anything they happen to fancy they will tenderly nurse, feed, put to bed, and discipline And when we reflect that a dog treats a scrap of wood as his prey, we can not regard as *a priori* out of the question for the animal's fostering instinct to fix on an object of the same kind. But while I admit this as an *a priori* possibility, I confess that I am unable to find an *a posteriori* experimental proof. The only cases that would serve in this connection that have come under my observation are in the report of the Loango Expedition. There Pechuël-Loesche says: "It

* Animal Intelligence, p. 495.

was something entirely new to me to see the monkeys take lifeless objects for playthings, and, like children, carefully put them to bed in their own sleeping boxes, as well as care for them during the day. Isabella devoted herself for a long time to a little canister, and Pavy to a little crooked stick of wood which in his wild capers he often hurled into the air. Once it flew too far and was appropriated by Jack, and thereupon deadly enmity ensued. As their chains were not long enough for them to reach one another, there was nothing for them to do but get as near together as possible and make horrible faces while they scolded. Their mutual hatred continued until I gave Pavy back his stick. Later he took to petting a musket ball, while Jack conceived a passion for a thermometer. As soon as he was free and not watched, he hurried to it and carried it off. He evidently delighted in the shining glass, but handled it so carefully that the instrument was never injured, even when he took it up in trees and on the roof and had to be coaxed down." *

It is questionable whether there is any analogy to play with dolls in such actions. At most, the putting things to bed and the care taken of the thermometer are all that could be considered so; but so long as better examples are wanting, these carry very little conviction.

But there is another and more important phase of the subject. When we see little girls playing foster mother to their younger brothers and sisters, and even to grown ones, and when we see, too, how lonely women lavish maternal care on lapdogs, which is really playing, we are not surprised at something of the same kind

* Loango Expedition, iii, p. 246.

among animals. Among the innumerable examples of the adoption of foster children and of animal friends, there are many that suggest play. I think, however, that those cases should be excepted where the mother, being robbed of her own young, has the young of some other animal thrust upon her by some experimenter—young which she regards with surprise but without a clear understanding of the deception.

We can as little speak of play in such a case as in that of a hen that tries to hatch marble eggs that have been placed under her. But there are still many cases that are like human play, and I will cite several such examples. The fact that the animal adopted is often maltreated and even in danger of its life does not argue against the playful intention. We see little girls frequently become very careless with their tenderly nurtured doll babies; we see them in the midst of maternal cares for an eatable toy make nothing of biting its head off; and we see the instinct of experimentation and destruction many times indulged even at the peril of their tame pets or little brother or sister, in spite of all the love for them.

Herr E. Duncker, of Berlin, observed, according to Büchner's report, a dog on a farm in Pyrmont, whose duty it was to watch the stock, and especially the poultry. He used to hunt up hidden eggs and bring them to the kitchen.* "One day he placed an egg on the sofa in the kitchen instead of on the stone floor, as usual. The little chick imprisoned in it was trying to break the shell, and after the egg was placed in a wadded

* L. Büchner, Liebe und Liebesleben in der Thierwelt, p. 185. Most of the observations cited here are from this book, which, it must be admitted, does not seem to be always of unimpeachable reliability.

basket the dog helped it out with his tongue and constituted himself its nurse. He let the chicken drink from the end of his tongue dipped in water, placed the basket in the sun, and petted and tended the little creature with unwearying care. When it grew up and was badly used by the other fowls he played protector, and the hen would fly on his back and appear to caress him"

Herr Wilibald Wulff relates that on a visit to the family of a friend in Schleswig he came upon a terrier lying in a basket holding two kittens with his fore paws, while two more clambered on his back. The lady of the house said, in answer to his questions, that he did this many times in a day—so often, indeed, that the old cat had deserted her young. He was far more careful of them than the mother herself, and would not allow any one to disturb the little ones. Dr. Matthes brought home a very young and helpless puppy, and noticed the next day that it had already been taken in charge by an old male dog. He lay down by the whining puppy, licked it, and growled at any one who came near. The following is related of a shepherd dog by Herr Heinrich Richter: "This remarkable and valuable dog had the habit that is common among good shepherd dogs of biting lightly the hind leg, just below the hock, of straying sheep. But he omitted to do this to one of the sheep, and only barked. Even at the command of the shepherd he refused to bite the sheep and only barked the more and licked it so that it became very bold, and allowed itself more freedom than ever. But woe to any other sheep who was emboldened by its example! He bit them all the more and punished them severely, as if to make up for his laxity in the other case. It was at last necessary to take away the favourite in order to prevent trouble, but even this was only a temporary remedy, for the dog turned his affection toward another sheep and acted as before." The owner of a truck farm, says the Revue d'Anthropologie, noticed that a basket which he had filled with carrots was unaccountably empty. The gardener, when questioned, knew nothing about it, and proposed hiding behind a lattice to watch for the thief. They did so, after refilling the basket. Soon a sound put them on their guard, and they saw the house dog take a carrot in his mouth and slink off toward the stable. Dogs do not eat raw carrots, so our watchers followed the rogue and discovered that he was taking the carrots to a horse in whose stall he was in the habit of sleeping. Wagging his tail, he presented his prize, and the horse naturally needed little urging to accept it. The angry gardener reached for a stick with which to punish the too zealous friend, but his master restrained him, and the scene was repeated until the supply of carrots was exhausted. This horse was evidently the dog's chosen favourite, for he scarcely noticed the other one that lived in the same stall, not to speak of giving him carrots. Fraulein Fanny Bezold, of Heidingsfeld, had a shaggy terrier named "Schnauz" that one day brought home a rabbit that he had caught at a farm about half a mile distant, and devoted himself to it. He played with his pet and defended it from the attacks of other animals and watched it anxiously when the children of the neighbourhood came in to see it. Herr Otmar Wild, in Zittau, writes to Büchner about the friendship between his setter one year old and a pullet They sleep side by side, or the hen on the dog's back. He expresses his tenderness by licking his little friend, and she shows her appreciation of it by picking about in his hair.

Many similar instances are recorded, tending to show that while this instinct is strongest in the female, it is not wanting in male animals, and that even among the fiercest animals the male assists in caring for and rearing the young.

Recorded examples are naturally most abundant among domestic animals. Mr. Oswald Fitch writes of a house cat: "It was observed to take some fish bones from the house to the garden and, being followed, was seen to have placed them in front of a miserably thin and evidently hungry stranger cat, who was devouring them; not satisfied with that, our cat returned, procured a fresh supply, and repeated its charitable offer, which was apparently as thankfully accepted. This act of benevolence over, our cat returned to its accustomed dining place, the scullery, and ate its own dinner off the remainder of the bones." *

If the playful character of this action seems doubtful, it is certainly present in Büchner's narrative which follows: At the mill near Hildburghausen there was a cat that rejoiced in the name of "Lies." She extended her maternal care not only to little chickens, but to young ducks and other birds as well. Once, immediately before the birth of four kittens, she brought six chicks, just hatched, to the basket prepared for her. She had some trouble in keeping the restless brood together, especially when her kittens came—in three days—but she never relaxed her care for the foster children. On the contrary, she soon brought to the nest three young ducks and a little red wagtail which she took from a nest near by. Her loving care was bestowed impar-

* Nature, April 9, 1883. See Romanes, Mental Evolution in Animals, p. 345.

tially on the motley crowd of nurslings, and she good-naturedly allowed the little chicks to peck at her nose and eyes. When they grew larger and ran about, they gave their foster mother endless trouble to bring them back and keep them safe, and by their constant pecking they made her neck quite bare. Fraulein Johanna Baltz, of Arnsberg, saw a large cat in the house of a friend acting as the protector of five little chickens, whose mother had been lost. The cat warmed and protected the tiny creatures when she first saw them, and it was a beautiful picture to see the cunning little heads, with their bright eyes peering out from under the gray fur.

Brehm has a great deal to say about this kind of play among monkeys. An orang-utang that Cuvier used to watch in Paris won the affection of two little kittens, which he often held under his arm or set on his head, although their sharp clinging claws must have hurt him. Once he examined their paws and tried to pull out the claws with his fingers. He did not succeed in this, but preferred to bear the pain rather than give up playing with his pets. A baboon named "Perro," that belonged to L Brehm, brother of the author of Thierleben, showed a strong partiality for young animals of all sorts. "When we were going to Alexandria we had him chained to the baggage wagon, giving him a long enough leash to do anything he wanted, short of running away. As we entered the city Perro spied a bitch lying in her kennel near the street and peacefully suckling four beautiful pups. To spring from the wagon and snatch one of the sucklings from its mother was the work of a moment, but regaining his place was another matter. The dog, enraged by the monkey's audacity, flew at him madly, and Perro had to exert all his strength to withstand her attack.

The wagon moved steadily onward, and he had none too much time to clamber in it when she sprang upon him. Holding the puppy between his upper arm and breast, and seizing the chain which impeded him with the same hand, he ran on his hind legs and defended himself bravely with one arm. His courageous defence won the admiration of the Arabs to such a degree that no one attempted to take the stolen puppy from him, and they finally drove the mother away. Unmolested, he brought the puppy with him to our stopping place, fondled it, nursed it, and cared for it tenderly, leaped over walls and rafters with the poor little creature, which seemed to have no taste for such exercises, left it in perilous places, and gave it privileges which might have been appreciated by young monkeys but were not agreeable to a dog. He was very fond of the little thing, but that did not hinder his eating all the food we brought it, actually holding it back while he robbed his innocent ward. I took the puppy away from him and sent it back to its mother that same evening." Another baboon that Brehm had behaved in the same way. "Atile loved pets of all kinds. Hassan, a long-tailed monkey, was the darling of her heart, so long as there was no question of eating. It seemed perfectly natural to her and no cause for gratitude that Hassan should share everything with her. She required slavish obedience of him, struck him on the mouth and emptied his plate without hesitation if he dared to think of enjoying anything alone. Her large heart was not satisfied with one pet, her love was all-embracing. She stole puppies and kittens whenever opportunity offered, and kept them for a long time. And she knew well how to render them harmless, for if they scratched her she would bite off their sharp claws."

"An interesting quality of our tame monkey," says Pechuël-Loesche, "is the way he has of choosing a particular animal, or even an inanimate thing, as the special object of his care. Strange friendships result from it. It is a familiar fact that apes often adopt the children of others of their own kind, care for them tenderly, and protect them to the last extremity. When our shepherd dog 'Trine' came home with her little ones, tormented by fleas, we placed them in the monkey house, where they were joyfully received and thoroughly cleansed, while the old dog looked on contentedly from outside. But a great commotion was raised when we attempted to take away the new pets, the monkeys evidently expecting to keep them. The good-natured ape Mohr formed a triple alliance with the gorilla and the ram Mfuka. Jack, the baboon, had a little pig for his friend, and often attempted to ride on its back. Later, in place of the cheerful pig, he had a half-grown dog for his chum, and they played together in the drollest manner. The morose Isabella chose a gray parrot for her pet, but the friendship was broken from the day she tried to pull out the parrot's beautiful tail feathers."

We will conclude with some examples from birds. "A friend of mine," says Wood, "has a gray parrot that is the tenderest and most devoted of foster mothers to any helpless little creatures. In the garden were a number of rosebushes surrounded by a wire fence thickly covered with vines. A pair of finches nested here, and were fed regularly by the people in the house, who were kindly disposed to all animals. Polly noticed these frequent visits to the rose garden, and the food scattered there, and determined to follow so good an example. Watching his chance, he escaped from the cage, imitated the call of the old finches, and carried to one after another of the young birds a bill full of his own food. But his manner was a little too brusque to suit the old birds, and they flew away in terror from the great gray stranger. Polly thus saw to his satisfaction the little ones orphaned and left entirely to his protecting care. From that time he refused to return to his cage, staying night and day with the foster children, and feeding them carefully till they were grown up. The little creatures would fly about and perch on his head and neck, and Polly would move very carefully and seriously with his burden." The naturalist Pietruvsky had a pond raven that always insisted on having company after a magpie was once accidentally placed in his cage. This companionship must have given him pleasure, for the next winter he chased any of the birds that came near when he was out of his cage. Tiring of the sport, he would catch a magpie and hold it in his claws, calling out until his attendant appeared, to put it in the cage. If the man dared to free the bird, the raven would keep on chasing magpies until he had his way; that accomplished, he would go into the cage of his own accord and there torment his beloved magpie, very much as monkeys tease their pets." *

Some birds that are reared with the first brood assist their parents in bringing up the second. "A family of swallows did this. Toussenel saw the first brood when they had hardly outgrown the nest themselves, lend a helping hand in feeding their little brothers and sisters." †

Altum assures us that the second brood of canary

---

* Büchner, Liebe und Liebesleben in der Thierwelt, p. 259.
† Ibid., p. 124.

birds is often fed by the first, and he has seen young kildees still in their first feathers bringing food to young cuckoos.*

If we now glance backward over the examples cited, it will be seen that the majority of them refer to abnormal conditions, like those in which the weaver bird displayed its skill. Most of the animals concerned had lost their own young and were trying to find an outlet for the fostering instincts already cited, and so a kind of make-believe was substituted for the natural expression of it, hence the origin of play. This is not quite the case when the animal adds strangers to its flourishing family, but it may be questioned even then whether the strange habit did not originate on some occasion when the animal could not exert its normal function. Yet I suppose those who regard the petting of dogs by lonely women as play may call this so too.

Play characteristics are, however, unmistakably present when experimentation and the desire for ownership are combined with the fostering instinct, and also when half-grown birds assist in caring for the younger ones. This latter seems to me the veritable play of young creatures, in which, however, imitation is perhaps as much involved as the nurturing instinct. It is certainly so in human play of this kind.†

7. *Imitative Play.*

I have already stated, in the previous chapter, that I subscribe to the views of those who, like Scheitlin, Schneider, Stricker, Wundt, and James, regard the imi-

* B. Altum, Der Vogel und sein Leben, p. 188.

† Perhaps the habit of many male birds of feeding their betrothed should also be reckoned among the indubitable plays.

tative impulse as instinctive, and now I must return to this vexed question. In order to get a clear view of the opposite theory, according to which imitation is of individual (not hereditary) origin, it is best to refer at once to the work which more than any other has influenced modern association psychology—James Mill's Analysis of the Phenomena of the Human Mind. So far as I can see there is nothing essential in later elucidations that is not contained in Chapter XXIV of this book. Mill proceeds from the assumption that the idea of a movement produces the impulse to perform the movement itself. The motion of swallowing furnishes a good example, for "if a friend assures you that you can not refrain for the space of a minute from this act, and you are tempted to try, you are almost sure to fail." Why is this true? Because directing the attention to the act of swallowing so strongly suggests the muscular feeling attending the act, that swallowing itself follows of its own accord. The same result follows when an idea of motion is suggested by the sight of it performed by another. For instance, there are certain feelings which we hardly notice accompanying gaping, and when we see another person gaping, we usually gape too; the act is so firmly associated with the accompanying feeling that the sight of the action arouses the feeling, which in turn calls for the act in ourselves. This explanation is expected to cover all phenomena included in the general name of imitation. It will be seen that the awakening of imitative impulse is here dependent on antecedent association. But for this it is necessary, as a rule, that the act in question shall have been repeated frequently, and thus, according to this theory, we would imitate only such motions as are already familiar. Were the associative connection between the "antecedent state of feeling" and the act itself firmly established by frequent repetition, the force of imitation would not be operative.

Against this definition the just and obvious objection is that imitation preferably selects what is new and unusual for its model, as the phenomena of fashion illustrate daily. When we see two people greet one another in the manner that we are accustomed to, we are not impelled to imitate them, though the associative connection is perfect. But if a leader of fashion displays a new way of lifting his hat, there are hundreds who can not resist the temptation to hold their hats, too, like warming pans, before them, or doing whatever the new mode demands. Further, this theory would make imitation a much stronger impulse in adults, whose associations are established, than in children, while the contrary is the fact. Nor does it explain any better the powerful influence of imitation in teaching the child new and unpractised movements of the limbs and vocal organs. Thus, when James Mill says—"All men have a greater or less propensity for imitation. This propensity is very strong in most children, and to it is due in large measure the rapidity with which they acquire many things, for example, the propensity to imitate sounds helps them to learn to talk quickly. . . . Children learn to stammer and to squint by imitating their companions, and we all know how common it is for young people to adopt the manners and expression of those with whom they associate"—he seems to me to prove by his own illustration that the exercise of imitative impulse does not use tracts learned by association, but rather inborn ones; in other words, that it is not acquired but inherited; it is an instinct. This is Herbert Spencer's view as it is set forth in the chapter on "Sociality and Sympathy" of his Principles of Psychology. He begins, it is true, with a purely associative principle in seeking to show how all the members of a herd of cattle often take to flight simultaneously, and how through the frequent repetition of this a strong association is gradually established between the signs of fright in another and the consciousness of fear, so that finally when only one animal perceives the danger, his fright is communicated to all the others. From this he goes on: "Evidently the process thus imitated must, by inheritance of the effects of habit, furthered by survival of the fittest, render organic a quick and complete sympathy of this simple kind. Eventually a mere hearing of the sound of alarm peculiar to the species will by itself arouse the emotion of alarm. For the meaning of this sound becomes known, not only in the way pointed out, but in another way. Each is conscious of the sound made by itself when in fear, and the hearing of a like sound, tending to recall the sound made by itself, tends to arouse the accompanying feeling. Hence the panics so conspicuous among gregarious creatures. Motions alone often suffice. A flock of birds, toward which a man approaches, will quietly watch for a while, but when one flies, those near it, excited by its movement of escape, fly also, and in a moment the rest are in the air. The same happens with sheep. Long they stand stupidly gazing, but when one runs all run, and so strong is the sympathetic tendency among them that they will usually go through the same movement at the same spot, leaping when there is nothing to leap over." *

* Herbert Spencer, The Principles of Psychology, p. 505. Audubon's description of the flight of passenger pigeons forms a parallel

I agree with Spencer in considering the imitative impulse hereditary, but must demur from his assumption of the inheritance of acquired characters, and take instead the principle of survival of the fittest, or selection, as the proper ground for a definition. In order to establish this connection, however, it must first be proved that imitation is useful, as I tried to do when I took the ground that it is an instinct which works directly toward the development of intelligence, since its tendency is to render many other instincts to a certain degree superfluous, and so encourage independence in the individual. This view is borne out by the fact that imitation is strongest in the more intelligent animals, such as highly developed birds and monkeys, and that man may be called the imitative animal *par excellence*.

Before I proceed to give instances of imitative play, it is necessary to point out briefly that this instinct is not by any means peculiar to gregarious animals, as seems to be the common impression. It is more or less operative in all the higher animals, especially while they are young. The family as well as the herd offers opportunity for its exercise, and we find examples of

---

to this observation of sheep: "It is impossible to describe the beauty of their aerial flight when a falcon attempts to snatch one from the flock. Startled, they fall back in a close mass, and then flow out like a living stream, pushing on with waving motion and in sharp angles, fall abruptly to the ground, and then mount straight upward in a column toward the sky, where they form a coiling line like a huge serpent. . . . It is remarkable how one flock after another will follow the same path. If, for instance, a bird of prey has disturbed one flock at a certain place, and they in consequence describe such angles, curves, and wavy lines, the next flock will do the same when it comes to that spot, as if it, too, had to escape from the fearful grasp of its enemy."

it among many animals that do not live in companies, as the instances which follow will show.

But when, it will be asked, can imitation be called play? Remembering our definition of play as instinctive activity exerted for purposes of practice or exercise, and without serious intent, it is easy to discriminate between imitation that is playful and imitation that is earnest. When a crow flies away with a warning cry, and the whole flock follows him, play has nothing to do with it. And the same is true of the beautiful instance given in Nature, September 12, 1889: "Two cats were on a roof, from which it was necessary to jump. Tom made the spring, but Tabby's courage failed and she drew back with a cry of distress, whereupon Tom leaped back, and, giving a cheerful mew as much as to say, 'See how easy it is,' jumped across again, followed this time by Tabby." But imitation appears in the character of play when young animals imitate the movements of their parents or other animals with no apparent aim but practice, when parrots reproduce every possible noise and tone, when monkeys copy their masters, and when animals have large gatherings for the purpose of competing with one another. Sully holds * that the imitative impulse is brought out only by such movements as are connected with "pleasurable interest"; but where movements of flight from approaching danger are concerned this can not invariably be true. Playful imitation, however, must always be connected with "pleasurable interest," and indeed it seems probable that such feelings of pleasure rest on the basis common to all play, which a searching examination will discover to be experimentation in this case, as

* The Human Mind, ii, p. 219.

14

well as in the others that we have considered. The delight in being able to say "I can," which we found in simple experimentation, becomes the joy of "I can too" in playful imitation, and under favourable conditions goes on to the pleasure of "I can do better" in rivalry.

Since playful imitation is often stigmatized as mimicry (Nachaffen), it seems peculiarly appropriate to begin with an example from the monkeys (Affen). The ancients were familiar with the imitativeness of monkeys, as their designations of them prove—the Greek μιμώ being one who imitates, and the Latin *simius* sounding much like *similis*.

The Egyptian word for monkey, though signifying rather the baboon in particular—an, anin, anan—likewise signifies imitator.* During the later Greek and Roman Empires monkeys were favourite pets because of their drollery. Their natural propensity was cultivated by teaching them all sorts of tricks, such as dancing, riding, driving a coach, playing the flute and the lyre. Ælian relates that monkeys had been known to scald little children in mistaken imitation of the nurse. According to Philostratus, who, it must be admitted, is not always trustworthy, the Indians employed monkeys in harvesting pepper. "They collected a small quantity of the fruit in a place prepared for it under a tree or at the foot of a hill, and then tossed it away as if worthless. The monkeys who attentively watched this proceeding came back at nightfall and, obeying their imitative impulse, made collections as the men had and left them. Next morning the Indians came and carried away the pepper thus harvested for them." †

---

* O. Keller, Thiere des classischen Alterthums, 1887, pp. 5, 323.

† Ibid., p. 4.

The probability of this story is indeed not enhanced when we learn that the method of catching the monkeys familiar to readers of Speckter's story book, namely, that of drawing on boots in their presence and then leaving the boots as a trap, was known to the ancients; but all such tales go to prove how impressive the monkey's imitativeness is. Modern accounts, too, are chiefly taken up with mimicry of human actions. Fr. Ellendorf relates a good example of curiosity, imitation, and experimentation combined in the person of a little black ape with a white head that he brought from Costa Rica. "On the first day that I let him run about in the sitting room, he sat before me on the table and eagerly examined everything there. Pretty soon he came upon a little matchbox and soon had it open and the contents scattered about on the table. I took a match, rubbed it on the box cover and held it near him. Full of astonishment, he rolled his little eyes and gazed at the clear flame. I struck a second and a third and held them out to him. At last he stretched out his paw, hesitatingly took the match, held it before his face, and watched the flame admiringly. Suddenly it touched his finger and he instantly threw the match away. I closed the box and placed it before me. From his hasty manner I thought he would open it at once, but he did not. He went near it, looked and smelt all about it without taking hold of it, then he came to me, making a low pleading sound and clinging to me, as if he were full of wonder and wanted to ask what this could be. Then he returned to the box, handled it all over and tried to open it, but could not do it alone, and came back to me with the same pleading tone. I struck another match on the cover and gave it to him when it was burned out. Then he took one and rubbed it on the cover and threw it away, but soon returned to it and, getting hold of it upside down, rubbed the wrong end. I turned it for him, and he struck it till it lighted. Now he was himself again, his whole manner showed the greatest joy and complaisance. He grasped the matches and struck at least a dozen." *

H. Leutemann contributes this about an orang-outang: "Most monkeys try to chew up whatever they can get at, and seem to take pleasure only (?) in destroying things, but ours, on the contrary, evidently tried to put to its proper use whatever was given to him. To my great surprise, he attempted to put on a pair of gloves, and, although he could not tell right from left, it proved that he knew what they were for. He supported himself on a light walking cane, and when it bent under him, made ridiculous motions to right it again." † Brehm tells of a chimpanzee: "After eating he at once begins to clean up. He holds a stick of wood in front of him or puts his hands in his master's slippers and slides about the room, then takes a cloth and scrubs the floor. Scouring, sweeping, and dusting are his favourite occupations, and when he once gets hold of the cloth he never wants to give it up."

The gorilla of which J. Falkenstein has given a detailed description was remarkable for his delicacy in eating: "He would take up a cup or glass with the greatest care, using both hands to carry it to his mouth, and set it down so carefully that I do not recall having lost a single piece of crockery through him, though we had never tried to teach him the use of such vessels,

* Thiercharaktere. No. 3, Affen. Gartenlaube. 1862, p. 87.

† H. Leutemann, Ein gebildeter Orang-Utang, Gartenläube, 1862.

wishing to bring him to Europe as nearly as possible in his natural condition." *

Romanes's sister has the following about the capucine ape already mentioned: "To-day he broke his chain . . . and got to the trunk where the nuts are kept, . . . and began picking at the lock with his fingers. I then gave him the key, and he tried for two full hours, without ceasing, to unlock the trunk. It was a very difficult lock to open, being slightly out of order, and requires the lid of the trunk to be pressed down before it would work, so I believe it was absolutely impossible for him to open it, but he found in time the right way to put the key in, and to turn it backward and forward, and after every attempt he pulled the lid upward to see if it were locked. That this was the result of observing people is obvious from the fact that after every time he put the key into the lock and failed to open the trunk, he passed the key round and round the outside of the lock several times. The explanation of this is that my mother's sight being bad, she often misses the lock in putting in the key; the monkey therefore evidently seems to think that this feeling round and round the lock with the key is in some way necessary to success in unlocking the lock, so that, although he could see perfectly well how to put the key in straight himself, he went through this useless operation first." †

Similar observations were made with two dogs, though imitation is nowhere so strong as with monkeys. Scheitlin describes his poodle's efforts at mimicry, which are in keeping with his remarkable intelligence. "He watches his master constantly, always no-

---

* Loango Expedition, ii, p. 152.

† Romanes, Animal Intelligence, p. 492.

ticing what he does; always ready to serve him, he is the right kind of eye-servant. If his master takes up a ninepin ball, he seizes one between his paws, gnaws at it, and is evidently annoyed that he can not take it up too. When his master looks for geological specimens, he hunts stones too, and digs with his paws when he sees digging going on. The master sits at a window admiring the view, the dog springs up on the bench near by, lays his paws on the window sill, and gazes, though not absorbedly, at the beauties of the scene. He always wants to carry a stick or basket when he sees his master or the cook carry one." *

There is probably something of playful imitation, too, in the howling of dogs when they hear music, for the dog which, for instance, accompanies the piano with mournful wails is often not compelled to listen to the music, but comes into the room voluntarily. I have said that I am doubtful whether the howling of dogs is always a sign of distress, and I am almost sure that it frequently is not when they howl to music; on the contrary, they seem to take pleasure in it. Moreover, there are cases on record where a rude attempt to imitate the music is apparent, though it is very easy to be mistaken about that. A friend of mine had, when he was a student, a female poodle named Rolla, with which he often gave performances for the entertainment of his friends. When he sang in a high falsetto voice the dog accompanied him with howls that unmistakably adapted themselves to the pitch of the notes. While there was, of

* Scheitlin. Thierseelenkunde, ii, p. 257. It will be noticed that not everything mentioned in this instance can be attributed to imitation. Opening doors is another instance of the same kind; there may be something imitative about it, but it is principally the result of effort to get out or in by scratching or pushing.

course, no such thing as following the tune, the impression was made on the hearer that the dog tried to sing with it, and was very proud of her skill. I should hesitate to relate this if others had not advanced the same belief. Scheitlin thinks that music may be painful to the dog, but goes on: "It may be questioned whether he does not, in his way, accompany it." * Romanes says the same thing: "With the exception of the singing ape (*Hylobates agilis*) there is no evidence of any mammal other than man having any delicate perception of pitch. I have, however, heard a terrier, which used to accompany a song by howling, follow the prolonged notes of the human voice with some approximation to unison; and Dr. Huggins, who has a good ear, tells me that his large mastiff, Kepler, used to do the same to prolonged notes sounded from an organ." †

Still more positive are some of the examples given by Alix; they really seem to border on the marvellous. "Père Pardies cites the case of two dogs that had been taught to sing, one of them taking a part with his master. Pierquin de Gembloux also speaks of a poodle that could run the scale in tune and sing very agreeably a fine composition of Mozart's (My Heart it sighs at Eve, etc.). It was called Capucin, and belonged to Habeneck, a theatrical director. All the scientists in Paris, according to the same authority, went to see the dog belonging to Dr. Bennati, and hear it sing the scale, which it could do perfectly. I myself know a poodle that accompanies his mistress very well when she plays the scale on the piano." ‡ Alix cites Leibnitz, too, who had seen a dog with such a capacity for imitation that he

* Thierseelenkunde, ii, p. 254.

† Romanes, Mental Evolution in Animals, p. 93.

‡ L'esprit de nos bêtes, p. 364.

could pronounce more than thirty words, making suitable answers to his master, and articulate clearly all the letters of the alphabet except M, N, and H.

The examples cited so far do not show us imitation in its real meaning; they are all the result of accidental offshoots from this powerful instinct, for the actual biological significance of imitative play is not expressed in movements or sounds that are unconnected with the struggle for life, but rather, to put it briefly, in playful self-discipline of young animals in the life habits of their kind It is sometimes very difficult to place the boundary between what is instinctive or hereditary and what is acquired by imitation. Still, it can hardly be questioned, after all that has been said, that imitative plays are an important adjunct to heredity during the youth of higher animals. The qualities of animals brought up by foster parents furnish a strong experimental proof of this. However the adopted animal may be limited in his development by inherited instinct, imitative impulse is still strong enough to bring about some startling modifications. I have not been able to collect many examples illustrative of this in mammals, the class to which I have hitherto confined myself.

Darwin tells us that "two species of wolves, which had been reared by dogs, learned to bark, as does sometimes the jackal," * and it seems quite certain that dogs brought up by cats learn many things from their foster parents. "From one account which I have read there is reason to believe that puppies nursed by cats sometimes learn to lick their feet, and thus to clean their faces; it is at least certain, as I hear from a perfectly trustworthy friend, that some dogs behave in

* Descent of Man, p. 42.

this manner." * Many such cases could be cited; indeed, Romanes found among Darwin's papers a manuscript of the late Professor Hoffmann, of Giessen, containing one.† But it is also a fact that dogs which have not been brought up with cats often have the habit of licking their paws and rubbing them over the face and ears, but no doubt the motions of the dogs in the other cases were noticeably like those of a cat. Imitation was more clearly displayed, however, by a King Charles spaniel mentioned in Miss Mitford's Life and Letters. This dog was suckled by a cat in its infancy at the home of Dr. Routh. He grew up with the horror of rain so characteristic of cats, and would not put his paw in a wet place; he would watch a mouse hole, too, for hours.‡ A certain Mr. Jeens also had a dog nursed by a cat, and it played with a mouse just as a cat does.#

Leaving such abnormal cases, I now pass on to consider the natural workings of this instinct. Every time a young animal imitates the movements of its elders without any aim beyond the unconscious one of practice, playful activity is indulged in. For example, I will relate what I have seen young polar bears do. There is a large flat stone in the bear pit, and the mother is constantly shoving it backward and forward. On one occasion it lay directly in her way, and she stepped over it, and the little one that was behind her, though he seldom cared to follow his mother about, tried to clamber over it too, and accomplished it with some difficulty. Brehm says that the only way he could get the young

* Descent of Man, p. 43.

† Romanes, Mental Evolution in Animals.

‡ Nature, May, 1873.

Ibid.

bears of the Hamburg gardens inside the inclosure of the bath was to run in himself, whereupon they all followed at once, otherwise their interest was absorbed by all sorts of things on the way.* This impulse to imitate motion may appear before the animal is able to distinguish between its mother and other objects, but simply follows anything that attracts its attention by moving—a clear proof that the impulse is hereditary. Hudson relates of young lambs that probably develop more slowly on account of domestication: "Its next important instinct (after sucking) which comes into play from the moment it can stand on its feet, impels it to follow after any object receding from it, and, on the other hand, to run from anything approaching it. If the dam turns round and approaches it, even from a very short distance, it will start back and run from her in fear, and will not understand her voice when she bleats to it. At the same time it will confidently follow after a man, horse, dog, or any other animal moving from it. . . . I have seen a lamb about two days old start up from sleep and immediately start off in pursuit of a puffball about as big as a man's head, carried past it over the smooth turf by the wind, and chase it for a distance of five hundred yards, until the dry ball was brought to a stop by a tuft of coarse grass. This blundering instinct is quickly laid aside when the lamb has learned to distinguish its dam from other objects, and its dam's voice from other sounds." † We often see among dogs how, when one goes over a ditch, his companions follow, and how the bark of one excites the rest at once. Wesley Mills emphasizes the

* Bilder aus dem Thiergarten in Hamburg, 2. Unsere Bären, Gartenlaube, 1884, p. 12.

† Hudson, The Naturalist in La Plata, p. 107.

extraordinary importance of imitation in development. He mentions the case of a mongrel dog that was placed with some St. Bernards when only twenty days old, and says: "One of the features of development greatly impressed on my mind . . . was the *influence of one on another* in all the lines of development. This was shown both negatively and positively in the case of the mongrel. After he began to mingle with the older dogs his progress was marvellous. He seemed in a few days to overtake himself, so to speak, and his advancement was literally by leaps and bounds." * The propensity of young bears to imitate their elders is often taken advantage of by tamers. Brehm gives an interesting description by K. Müller of the education of young stone martens: "The mother is most attentive to the exercising of her young, as I have had occasion to notice several times. In the park a wall five metres high is connected with the shed where a pair of martens with four young ones are housed. At daybreak the mother crept out cautiously, stealing like a cat some distance along the wall and crouched there, quietly waiting. There the father joined her, but it was some moments before the young ones came out. When they were all together the parents rose and in five or six bounds covered a considerable stretch of the wall and vanished, and I heard, though it was scarcely audible, the sound of a spring into the garden. The little ones followed with hurried leaps and climbed up on the wall with the aid of a poplar tree growing near. Hardly had they reached their parents when the latter sprang away again, this time to a lilac bush, and now the young ones followed them without hesitation. It was aston-

* *Loc. cit.*, part iii, p. 219.

ishing how by a hasty glance they could detect the best route And now began the running and leaping with such zeal and at such a breakneck speed that the play of cats and foxes seemed mere child's play beside it With every moment the pupils grew more agile, rushing up and down trees, over roofs and walls with a rapidity that proved how necessary it was for the birds of the garden to be on their guard."

Turning now to birds, I begin where we left the four-footed animals, for among the former imitation of parents is much more the rule than among mammals, and especially so with singing birds. I should like to call attention, in this connection, to the position of Wallace, who, though he found that the facts did not bear him out in the attempt to refer everything to imitation, has still given us some valuable reflections on its pedagogical aspects. There are cases on record of birds which have been reared apart from any of their species and never learned their characteristic song perfectly, while on the strength of other observations it seems just as certain that instinct alone is sufficient to teach them not only simple calls, but genuine song. Romanes's conclusion seems to be the right one—namely, that song and the other general capacities of birds are instinctive, but can never be so quickly nor so perfectly expressed as when the parents serve as models.* That the value of imitation is not to be despised is seen in the many cases where young birds are brought up by some other kind, whose song they adopt, showing that their imitative impulse is stronger than the hereditary disposition to the song of their own kind. We are again in-

* Weinland, too, reached essentially the same conclusion after years of experience (Der zoologische Garten, III, 1862).

debted to Weinland's diary for the records of a family of canaries. On May 14, 1861, the shells were broken. One bird with a black head was the strongest and most active of the brood. On June 2d little Black-head sang for the first time, or rather he twittered while his father sang. This is a good example of playful imitation.*

In Thuringia chaffinches are bred that have a specially acquired song, no one knows why, probably through unconscious selection. If young ones are reared near those having the special song, they catch the note in their play.†

The many cases, too, where the female imperfectly imitates the song of the male may be playful. Then there is the well-known tendency of song birds to make themselves heard when another is singing, when a piano is being played, or conversation carried on. Imitation here becomes rivalry.

But imitation is not confined to singing in young birds. "They are like little monkeys," says Hermann Müller; "example always excites them. When one little one, whose wings are feathered or not, as the case may be, begins to flutter, all the little wings are agitated." This observation seems to prove that it is not individual experience alone that causes a flock of grown birds to take flight simultaneously ‡ I have already pointed out that young chickens take twice as long to learn to walk alone as when they have the maternal example before them, and that waterfowls go into the water with their young and swim before them. Darwin says, in his manuscript left unpublished: "It might have been

---

* F. Weinland. Eine Vogelfamilie, Der zoologische Garten, 1861.

† Naumann, Naturgeschichte der Vögel Deutschlands, iv, p. 27.

‡ Büchner, Aus dem Geistesleben der Thiere, p. 30.

thought that the manner in which fowls drink, by filling their beaks, lifting up their heads, and allowing the water to run down by its gravity, would have been specially taught by instinct; but this is not so, for I was most positively assured that the chickens of a brood reared by themselves generally required their beaks to be pressed into a trough, but if there were older chickens present, who had learned to drink, the younger ones imitated their movements, and thus acquired the art"*

It is probable that the imitative impulse comes into play in similar fashion many times in an animal's life, when we are entirely unable to prove its presence or influence.

The imitation by birds of the songs of other species is very common. It would be an endless task to cite even a portion of such descriptions as are found, for example, in the works of Naumann, Beckstein, Russ, the two Brehms, the Mullers, etc. I therefore confine myself to a choice among examples where the imitative impulse appears in greatest perfection, where not only bird voices but those of men, as well as sounds like the creaking of doors or a mill wheel, playing on pipes, and spoken words are faithfully copied It should be noted that this strange habit is not peculiar to birds which lack a song of their own,† such as parrots and the crow family, but appears in good singers as well. The wild canary, which has a great talent for mimicking other

---

* Romanes. Mental Evolution in Animals, p. 229. [Lloyd Morgan's more recent experiments (Habit and Instinct) confirm this but go to show that after the wetting of the beak the young bird throws up the head and swallows instinctively.]

† As Romanes seems to think, Mental Evolution in Animals, p. 222.

birds,* when tamed can be taught to speak; and the American mocking bird, which Dr. Golz, of Berlin, a most competent judge, gives the precedence over all species of nightingales.† imitates everything conceivable, even to the creaking of a rusty hinge.‡ I think this is easy to explain: the singers have had their powers improved by practice in learning their complicated songs, and parrots and crows are endowed with unusual ability for speech, for which imitation is particularly essential. According to Karl Russ, these birds manifest a certain degree of comprehension of the meaning of words uttered by them, while other talking birds babble meaninglessly, or warble the words in song. I take a canary for our first example. Karl Russ says: "On the 23d of April, 1883, I called on the wife of Commissioner Graber in Berlin to see and hear her little feathered talker. The lady received me with the warning that I had probably come in vain, for the bird did not seem inclined to talk that day. She told me that she had had him for about three years, and believed him to be quite young. From being a fine singer he suddenly stopped, probably as a result of moulting, and as his silence continued for some time she frequently said to him, 'Sing doch, sing doch, mein Mätzchen, wie singst du? widewidewitt!' 'You can imagine my amazement,' she continued, 'when the canary pronounced for the first time the words I had thus quite accidentally said

---

* Karl Russ, Handbuch für Vögelliebhaber, Züchter, und Händler. ii. p. 130.

† Ibid., i, p. 284.

‡ [This the present editor can confirm in the greatest variety of detail. He has heard two of these birds together imitating the "clipping" of a gardener's trimming-shears, as if competing with each other.]

to him I hardly trusted my senses and could not understand it at first.' When the lady had told me this, she turned to the canary and repeated the same words. He began to twitter, and in the midst of his song we heard 'Widewidewitt! wie singst du, mein Matzchen? singe, singe Mätzchen, widewidewitt!' Again and again he repeated it, and the words became clearer and plainer. The bird did not articulate the words in human tones, but wove them into his song. The sound was always harmonious, and from the first one could understand the words, but they became more distinct as one listened." * Russ quotes the report of Mr. S. Leigh Lotheby, in the Proceedings of the Zoölogical Society of London for 1858. A canary bird was brought up by hand and his first song was very different from the characteristic one of his kind.† He was constantly talked to, and one day when he was about three months old he astonished his mistress by pronouncing after her the caressing words that she used to him, "Kissie, kissie," and then produced the smacking sound of a kiss. From time to time the little bird learned other words, and amused his friends by his manner of using them for hours at a time (except when moulting) in various combinations according to his fancy, and as clearly as the human voice can produce them: "Dear, sweet Fitchie, kiss Minnie, kiss me then, dear Minnie, sweet, pretty little Fitchie, kissie, kissie, kissie, dear Fitchie, Fitchie, wee, gee, gee, gee Fitchie, Fitchie." The habitual song of this bird was more like that of a nightingale, and the sound of a dog whistle used in the house was often heard in it. He also whis-

* Karl Russ, Allerlei sprechendes gefiedertes Volk, 1889, p. 169.

† Another proof of the great importance of imitation.

tled very clearly the first strain of "God save the Queen." *

The European bullfinch, whose natural song the Thuringians call "rolling a wheelbarrow," though it has great variety, readily learns to whistle songs. The elder Brehm says of it: "I have heard the red linnet and the black thrush whistle many tunes not badly, but no other bird attains a purity, softness, and richness of tone equal to the bullfinch. It is incredible how far he can be trained. He often learns the melody of whole songs and produces them with such a flutelike tone that one never tires of hearing him." Herr Theodor Franck, of Berlin, writes that his bullfinch was quite a skilful whistler. "But the accomplishment that endeared him to us is his having learned to repeat the words that my wife and I address to him as he hangs in our chamber. 'Little man, are you there?' or 'Courage, Mannikin, courage.' The red linnet has a wonderful facility in imitating the songs of strange birds, as well as real melodies and discords. The crested lark sometimes learns as many as four different tunes, and mimics birds and animals as well.

Count Gourcy writes to the elder Brehm of the bunting of southern Europe: "Its call resembles, in all but one deep tone, the decoy cry of the crested lark. Its song is magnificent, and really extraordinary for its variety. It possesses the rare power of changing the quality of its voice at will, producing now high, shrill notes and then tones so clear as to astonish the hearer. Usually some strains of the nightingale's song follow the first call, then comes the long-drawn, deep cry of the blackbird, in which the familiar 'Tack, tack'

* Russ, *loc. cit.*, p. 174.

is sounded very beautifully. After this follow strains that sometimes include the whole song of the chimney swallow, song thrush, quail, woodlark, linnet, field lark, and crested lark, the finch and sparrow, the laughter of woodpeckers, and shricking of herons, all of which are produced in the natural tone."

"The paradise bird," says Alix, "has equally with the group of singing birds excellent imitative powers. I had one, writes Blythe, that mimicked the *kittacincla macrowra* so well that no one could distinguish their songs. I also owned another having the same power. There is no sound that it can not imitate. It crows so perfectly that cocks answer it, and it barks and mews quite as well, bleats like a goat or sheep, howls plaintively like a beaten cur, croaks like a crow, and sings the song of many birds." * The American mocking bird, which has been referred to as a splendid singer, has also a remarkable talent for mimicry. "In its native woods," says Brehm, "it mocks the wild birds; near human dwellings, it weaves into its song all sorts of sounds heard there Crowing, cackling, quacking, mewing, barking, creaking of doors and weathervanes, the hum of a saw and rattle of a mill—all these and a hundred other noises are reproduced with the utmost faithfulness." † European thrushes, too, have, Brehm says, a strong propensity to imitation, though they confine it more to their own kind. Yet the blackbird "mimics birds of strange species and sometimes becomes a veritable mocking bird." ‡ According to

* F. Alix, L'esprit de nos bêtes, p. 362.

† See, too, Hudson's beautiful description of Patagonian mocking birds. The Naturalist in La Plata, p. 276.

‡ Romanes says that both the blackbird and the crow have been known to mimic a cock. Mental Evolution in Animals, p. 242.

Brehm, too, the stone thrush and blackbird are talking birds as well, though Russ questions this. Beckstein has shown by experiment that the stone thrush can be taught to whistle melodies. The natural song of the starling consists in complicated "fluting, piping, twittering, and chuckling sounds." * But they copy the songs of other birds, cock crows, hen cackles, door creaking, etc., and have been known to attempt human speech. The older writers have no doubt exaggerated this capacity, but the following testimony will show how far the starling can be educated. K. Dittman writes of the learned starling owned by the master shoemaker G. Dom: "The bird learned with surprising ease to whistle the 'Call of the Fire Brigade' and other tunes. His name was Hans, and his master would call out often during a lesson, 'Careful, Hans, careful'; he quickly learned this and pronounced the words with perfect ease, proving his ability to talk as well as catch a tune. It was very comical to see him stand among the cobblers and call out, 'Hurrah for Bismarck!' or cry 'Pickpocket!' when any one came in the door." Another starling could say all the following: "Have you heard the news? My, but it's good! Good morning; are you up already? What do you know that's nice? How is the Kaiser getting on? And what's the matter with Bismarck? God bless you! Are you there? Take a seat; are you a fool? Yes, yes!" † But the Asiatic magpie is the most talented of all the starlings, and claims among its connections some of the very finest singers.‡

Passing by many other imitative birds, I turn now to

* Russ, Allerlei sprechendes gefiedertes Volk, p. 138.

† Ibid., p. 145. ‡ Ibid., p. 160.

the ravens. Dickens's description of one in the preface to Barnaby Rudge is too familiar to need quoting for English and American readers

Naumann's remark that ravens are more easily taught to speak than parrots is probably an exaggeration,* but it is undeniable that imitativeness has reached an extraordinary development in these birds. Chr. L. Brehm says of one: "His talent for mimicking every sound with his voice is remarkable. He laughs like the children, coos like the pigeons, barks like a dog, and talks like a man. His reproduction of certain tones is so deceptive that some of my friends, hearing him for the first time, could not be convinced that such sounds actually proceeded from a bird 'James, come here,' 'Rudolph, come in,' 'Don't you hear me, Christine?' and much more, he articulated perfectly and voluntarily, not because it was required of him He picked up all these words, for no one ever took the least pains with him, but he could be heard trying new words every day, of those that he constantly heard around him." † The Müllers, too, mention similar instances.‡

But of all birds, parrots are the ones that manifest playful imitation most strongly. Their powers were well known as far back as the Romans, for Cato thunders against the luxuriousness of the *jeunesse dorée* of his time for flaunting in the streets with parrots on their thumbs; and courtiers under the emperors taught the birds the formula of greeting and gratulation to the

* Naturgeschichte der Vögel Deutschlands, ii, p. 47.

† Beiträge zur Vögelkunde, ii, p. 80.

‡ A. and K. Müller. Wohnungen, Leben und Eigenthumlichkeiten in der höheren Thierwelt, p. 364.

Cæsar. Kristan von Hamle, one of the lesser Thuringian minnesingers, expressed the wish in 1225:

> "Oh, that the green grass too could speak
> As doth the parrot in his cage!"

And Celius tells us that the parrot belonging to Cardinal Ascanius could recite the twelve articles of faith.*

This highly developed impulse of imitation in the parrot is probably due to the unusual intricacy of their native language. Marshall says: "One must hear them when they do not know that they are observed, and when a pair chat together, to appreciate their fulness of tone and the variety of meaning they can convey in one of their long conversations." † To learn speech so complicated as this requires imitative power, and in this case it seems especially developed in the imitation of sounds.‡

In selecting some examples to insert here I regret being obliged to omit a very remarkable one related by Brehm. In his battle for individual reason against instinct he became strangely credulous, and all his examples bearing on that topic are under the shadow of that imputation. The following collection, however, is vouched for as unimpeachable by Karl Russ, in his Feathered World. Of the wonderful gray parrot belonging to Director Kastner in Vienna it is said: "For a while after coming to us he spoke only when alone in the room, but soon took to chattering without noticing his surroundings, joining heartily in a laugh,

* W. Marshall, Die Papageien, Leipsic, 1889, p. 3.

† Ibid., p. 42.

‡ It is worthy of note that I have not been able to find a single instance of imitation of the speech of other beings, either man or animal, by a monkey; and yet many kinds have a well-developed language of their own.

too, on occasion. On hearing a low whistle, he said, 'Karo, where is Karo?' and himself whistled for the dog. He could whistle with rare skill a great variety of melodies, and reproduce any air perfectly. As soon as the dinner bell rang he called the waitress louder and louder until she appeared. If a knock came at the door, he said 'Come in,' but was never deceived by any one in the room. If he saw preparation made for uncorking a bottle, he made the noise long before the cork was out. He talked to himself in soft, gentle tones, "You good, good Jacky,' etc., but would call out in a strong masculine voice, 'Turn out, guard!' etc., and make the roll of a drum. He could count, and if he made a mistake or mispronounced a word he would go back and try it again till it was all right. When the green parrot standing near him screamed, he first tried to quiet her with a reproving 'Pst!' but if that did not avail he called out in a loud voice, 'Hush, hush, you!' He loved to talk to himself late in the evening, and regularly closed his monologue with the words, 'Good night, good night, Jacky.'" *

Herr Ch. Schwendt says of his gray parrot: "My parrot is a living proof that one should never despair of teaching these birds to speak. I had to wait eight months before he brought out the word 'Jacob,' but the ice once broken, I was richly rewarded for my patience; he learned something new almost every day, and now after four years he knows more than I can tell. There is hardly any expression commonly used in the family that he has not learned to repeat, and how well he knows how to apply them! He speaks of everybody in the house and all the animals by name, whistles to the

* K. Russ, Die sprechenden Papageien, 1887, p. 28.

dogs and orders them about, coaxes the cats or scolds them. He has the names of all the other birds at his tongue's end, and answers with the right one at the sound of their voices, never confusing them. He can alter his voice from the tenderest caressing tone to a gruff command, 'Present arms!' or the like, all in tones astonishingly human and with clear pronunciation. He recites verses and praises himself when he has not made any mistakes; but if he does, he says, 'That's not it, stupid!' He uses every greeting at the right time of day, and can apply everything he knows with propriety. He can count correctly up to eight." * Such examples shows that with parrots something more than mere blind imitation is involved, since such highly endowed specimens as this one can make the proper connection between the acoustic symbol and its mental import, but great caution must be exercised to avoid exaggerated interpretations of their performances. The gray parrot of the African traveller Soyaux showed a greater ability to learn: "An old bird when caught, he never was thoroughly tamed, but was greatly admired on account of his size. He talked very little, only rarely pronouncing the word 'kusu,' which is the native designation for parrots, but his great forte was whistling, in which I have never seen him excelled. Not that he was so specially skilful in whistling whole songs, but the modulation was wonderful—as strong, full, and clear as a bell, like high organ notes. He would roll up and down the scale, skipping a note and sounding it after the succeeding one His memory of African bird notes was remarkable, and he imitated perfectly the call of plovers, cranes, etc." †

* K. Russ, Die sprechenden Papageien, p. 29. † Ibid., p. 31.

Cockatoos, ring parrots, and some other varieties also learn to speak readily, the latter having been known to acquire as many as a hundred words in various languages, and articulate them perfectly. The cockatoo is a very sociable bird, and indulges in much gesticulation and genuflection while speaking. "Nodding the head and making the drollest bows that shake his bright crest, he turns and clambers about and laughs with real appreciation of the joke when he mimics the movements, words, or cries of another."*

In concluding this series of examples I wish to include a few illustrating more directly the social aspect of imitation. I remember that Spencer says it is "sympathy" that induces a whole flock of birds to rise when one flies off, and I think that such effects of imitation on masses may at times be playful as well. The following interesting remark of James's will serve to illustrate what I mean: "There is another sort of human play, into which higher æsthetic feelings enter. I refer to the love of festivities, ceremonies, and ordeals, etc., which seems to be universal in our species. The lowest savages have their dances more or less formally conducted. The various religions have their solemn rites and exercises, and civic and military powers symbolize their grandeur by processions and celebrations of divers sorts. We have our operas and parties and masquerades. An element common to all these ceremonial games, as they are called, is the excitement of concerted action, as one of an organized crowd. The same acts, performed with a crowd, seem to mean vastly more than when performed alone. A walk with the people on a holiday afternoon, an excursion to drink

* K. Russ, Die sprechenden Papageien, p. 117.

beer or coffee at a popular 'resort,' or an ordinary ball-room, are examples of this. Not only are we amused at seeing so many strangers, but there is a distinct stimulation at feeling our share in their collective life. The perception of them is the stimulus, and our reaction upon it is our tendency to join them and do what they are doing, and our unwillingness to be the first to leave off and go home alone." *

From the last words it is evident that such mass plays are based on imitation, and that social influences of the greatest importance belong to them. G. Tarde regards it as *the* fundamental principle of all society. There are, he says, in his daring way of drawing analogies, three great laws of repetition: undulation in physics, the nutritive-generative principle in physiology, and imitation in psychology. Imitation makes society: "*la société c'est l'imitation.*" †

Since, then, imitation has so much to do with the social life of men and animals, we are not surprised to find it prominent in their sports. Herds and flocks unite in various games, vocal practice, and even in trying the arts of courtship and combat, when the playful

* W. James, The Principles of Psychology, ii, p. 428.

† G. Tarde, Qu'est-ce qu'une société? Revue philosophique, xviii (1884). See the article on Imitation by J. Mark Baldwin (Mind, 1894), who regards the change produced by expansion and contraction in protoplasm as the first manifestation of organic reactions of the imitative or "circular" type which therefore becomes a central phenomenon of life. Professor Baldwin has now developed his psychological theory of imitation in his work, Mental Development in the Child and the Race, to which I have already frequently referred. And in his later work (1897), Social and Ethical Interpretations in Mental Development, he shows that the sense of self upon which social organization rests is developed only by imitation.

act of one animal spreads through the whole company like a sudden contagion. Very often, and especially in the courtship plays, what is at first taken up in a mere spirit of imitation becomes the sharpest rivalry.

It is difficult to speak with assurance in this matter, of the larger mammals especially, but I have no doubt whatever that the mad rushing of great herds of wild horses, deer, and goats that is so common on the plains is as often the result of a general desire to play as of apprehended danger. When one cow in a herd leaps down the slope where they are grazing, a large part of the herd will often follow with sportive bounds and mock fighting. Even a drove of pigs will show playful movements that are infectious; the wild gambols of seals and dolphins have already been instanced. Hudson saw a very beautiful game played by a number of weasels. "They were of the common larger kind of weasel (*Galictis barbara*), about the size of cats, and engaged in a performance that suggested dancing, which so absorbed their attention that they did not notice me when I came within four or five metres of them to see what they were doing. It proved to be a chase on a deserted viscacha mound; they all, about a dozen in number, ran swiftly across, jumping over the holes, turned at the end of the mound and came flying back without ever colliding with one another, though they were apparently beside themselves with excitement, and their paths crossed at every possible angle. It was all done so quickly and with such constant changing of direction that I found it impossible to follow a single animal with my eye, however hard I tried." *

* The Naturalist in La Plata, p. 384.

If the destructive impulse that seizes children in the presence of beetles and frogs, or even larger animals, such as cats, has anything playful about it, an example that Hudson relates in his chapter on "Some Strange Instincts of Cattle" may not be out of place here. This execution, as it were, of sick or wounded companions is also common among birds and carnivorous animals that live in companies. When a rat is wounded, his comrades slay him, indeed, Azara says that pinching the tail of a captive rat until he squeals is enough to make his companions fall upon him and bite him to death.* Hudson, speaking of his childish memories, says: "It was on a summer's evening and I was out by myself at some distance from the house, playing about the high exposed roots of some old trees; on the other side of the trees the cattle, just returned from pasture, were gathered on the bare, level ground. Hearing a great commotion among them, I climbed on one of the high exposed roots and, looking over, saw a cow on the ground, apparently unable to rise, moaning and bellowing in a distressed way, while a number of her companions were crowding round and goring her." † To the same category belongs Dr. Edmonson's somewhat fantastic description of an execution by crows. "In the northern parts of Scotland and in the Faroe Islands extraordinary meetings of crows are occasionally known to occur. They collect in great numbers as if they had been summoned for the occasion; a few of the flock sit with drooping heads and others seem as grave as judges, while others, again, are exceedingly active and noisy; in the course of about one hour they disperse, and it is not uncommon after they have flown away to find one

* The Naturalist in La Plata, p. 343. † Ibid., p. 339.

or two left dead on the spot." * Hudson explains such instances of frantic murder as the first two as caused by the impulse to relieve tortured comrades—the enraged animals make for the enemy that has caused their distress, and in a kind of madness fall upon his victim, to whose rescue they have come This does not seem plausible to me. Darwin and Romanes are of the opinion that it is a special instinct, useful to the species; but this also seems to me to be an inadequate explanation, for it does not tell us why it is not enough for the herd simply to abandon such unfortunates to their fate. The truth of the matter is, I think, that we have here no special instinct, but another form of the old impulse for fighting and destroying that is always ready to break out. "In the misfortune of our best friends there is always something pleasurable," say La Rochefoucauld and Kant. The sight of a cripple or an intoxicated person often arouses in children and savages a wild desire to worry and torment, and just so the inherited impulse to injure and destroy finds expression in the animal and is communicated by means of the powerful principle of imitation, through a whole herd, before quite peaceable. Actual play it can not be said to be, and therefore I shall not spend any more time over the question, though, in a certain sense, it resembles play.

We do find genuine play in the vocal practice that so many mammals constantly indulge in. A zoölogical garden where several lions are kept is a good place to observe this. I have often listened while a young lion lifted up his voice, at first with a peculiar gurgling sound, then in thundering roars in which others joined

* Romanes, Animal Intelligence, p. 324.

in a frightful concert that made the whole house tremble. Brehm says: "Lions in the wilderness, too, delight in this; as soon as one lifts up his mighty voice all others within hearing join him, making magnificent music in the primeval forest." Most remarkable are the concerts of howling apes, whose din fills the South American wilderness; with them, too, a solitary voice is heard at first which incites the rest to accompany the leader. I believe this is a phenomenon of courtship, like the nocturnal wailings of cats.

Imitation seems to be even more provocative of concerts among the birds. I included under experimentation descriptions of the chakar, of the familiar cries and gabbling of geese, ducks, and crows, and of the myriad-voiced concerts of our woodland singers which mutually incite one another. I cite a description of Hudson's which might apply to many birds that delight our eyes by their evolutions in flight. "In clear weather they often rise to a great height and float for hours in the same neighbourhood—a beautiful cloud of birds that does not change its form, . . . but in this apparent vagueness there is perfect order, and among all those hundreds of swiftly gliding forms each knows its place so well that no two ever touch; . . . there is such wonderful precision in the endless curves made by each single bird that an observer can lie on his back for an hour watching this mysterious cloud dance in the open without tiring."

The black-headed ibis of Patagonia, which is almost as large as a turkey, carries on a strange wild game in the evening. A whole flock seems to be suddenly crazed; sometimes they fly up into the air with startling suddenness, move about in a most erratic way, and as they near the ground start up again and so repeat the game, while the air for kilometres around vibrates with their harsh, metallic cries. Most ducks confine their play to mock battles on the water, but the beautiful whistling duck of La Plata conducts them on the wing as well. From ten to twenty of them rise in the air until they appear like a tiny speck, or entirely disappear. At this great height they often remain for hours in one place, slowly separating and coming together again, while the high, clear whistle of the male blends admirably with the female's deeper, measured note, and when they approach they strike one another so powerfully with their wings that the sound, which is like hand-clapping, remains audible when the birds are out of sight. The most beautiful member of the quail family found in La Plata is the ypecaha—a fine, strong bird about the size of a hen. A number of them choose a rendezvous near the water. One raises a loud cry three times from the reeds near this spot, and the invitation is quickly responded to by the other birds, who hasten thither from every direction till ten or twenty are collected. Then the performance, which consists in a frightful concert of screams, begins in tones that are strongly suggestive of the human voice when it expresses extreme terror or agonizing pain. A long, penetrating cry of astonishing force and violence follows the deeper tones as though the creature would exhaust all its strength in this alarm. Sometimes this double call is repeated and is accompanied by other sounds that resemble half-smothered groans, and all the while the birds run about as if possessed, their wings outstretched, their beaks wide open and held up. After two or three minutes the company quietly breaks up.

Jacanas, strange birds with peculiar cockscomblike head decorations, spurs on their wings, and long, thin claws, give a kind of exhibition which apparently serves the purpose of displaying their wing decorations, which are concealed, under ordinary circumstances. From twelve to fifteen of them come together at the signal, form in a close mass, and, while producing short, quickly repeated notes, unfurl their wings like a standard of banners. Some hold them upright and rigid, others keep them half open with quick vibration, and still others wave them with slow, regular motion back and forth.*

In all these examples, which might easily be multiplied, courtship is evidently the unconscious basis, as any unbiased mind must be convinced by a glance at the following chapter. When the contagious influence of imitation becomes a factor in mass games, they are easily converted into veritable orgies. I think we encounter here among the birds the same principles that govern ethnology and the history of human civilization. Their plays correspond with our general dance that is so closely connected with sexual excitement, and the examples given above may be likened to Middendorf's description of a dance of savages. "The dance soon became boisterous, the movements mere leaps and hops, the faces inflamed, the cries more and more ecstatic as each tried to exceed the others. The fur coats and breeches were thrown off, and they all seemed to be seized with a frenzy. Some, indeed, made an effort to withstand it, but soon their heads took the motion, now right, now left, till suddenly the onlookers leaped among the dancers as if they had broken some controlling bonds, and widened the circle." †

* See The Naturalist in La Plata, p. 265.

† O. Stoll, Suggestion und Hypnotismus in der Völkerpsychologie, Leipzig, 1894, p. 24.

The principal difference is that the motions of the human dancer less clearly betray the courting instinct, though it is none the less there, however latent, and we may learn much from the courtship of birds that is applicable to man as well.

8. *Curiosity.*

Curiosity is the only purely intellectual form of playfulness that I have encountered in the animal world. It is apparently a special form of experimentation, and its psychologic accompaniment is attention, which indeed is a requisite to the exercise of most of the important instincts. Leroy has said that three things demand the animal's attention: the cravings of hunger, those of desire, and the necessity of avoiding danger,* and Ribot, too, assigns the same reasons for its importance.† This important faculty finds a playful expression in curiosity, which may be called sportive apperception. This function, that forms an essential element in the activity of all the principal instincts, especially those of feeding and flight, oversteps its utilitarian character in curiosity and becomes play. The necessity for mental exercise is the primary reason for this kind of playfulness, added to the increase of knowledge. As James expresses it, it aids in the preservation of the species, "inasmuch as the new object *may* always be advantageous" ‡

* Lettres philosophique sur l'intelligence et la perfectibilité des animaux, p 71.

† Th. Ribot, Psychologie de l'attention, p. 44 I think Ribot is right in emphasizing hunger and fear more than desire.

‡ W. James, Principles of Psychology, ii, p 429. I have already pointed out that all play employs the attention, and, indeed, all the mental powers. Sikorski shows that attention is developed in

Next to the child, the monkey is the most curious of animals. I repeat the anecdote often cited from Darwin as the best example we have: "Brehm gives a curious account of the instinctive dread which his monkeys exhibited toward snakes, but their curiosity was so great that they could not desist from occasionally satiating their horror in the most human fashion—by lifting up the lid of the box in which the snakes were kept. I was so much surprised at his account that I took a stuffed and coiled-up snake into the monkey house at the Zoölogical Gardens, and the excitement thus caused was one of the most curious spectacles which I ever beheld. Three species of the Cercopithecus were the most alarmed; they dashed about their cages and uttered sharp signal cries of danger which were understood by the other monkeys. . . . I then placed the stuffed specimen on the ground in one of the larger compartments After a time all the monkeys collected round it in a large circle, and, staring intently, presented a most ludicrous appearance. . . . I then placed a live snake in a paper bag, with the mouth loosely closed, in one of the larger compartments. Then I witnessed what Brehm has described, for monkey after monkey, with head raised high and turned on one side, could not resist taking momentary peeps into the upright bag at the dreadful object lying quiet at the bottom " *

That dogs, too, are curious is a familiar fact. A strange dog attracts immediate attention, and a favourite curb excites as much interest as a lonely tourist bestows on the register of his inn. Curiosity adds to the watch-

---

children by play (Revue philosophique, April, 1885). But, in curiosity, attention itself becomes play.

* Descent of Man, vol. i, p. 41.

dog's value by inciting him to investigate every sound Scheitlin, overlooking the monkey, calls dogs the most curious of animals next to goats, and, strange to say, nightingales.* The curiosity of a dog is very ludicrous when a beetle runs before him; evidently he is a little afraid of the tiny creature, but he can not rest until he has smelled it all over. A dog that Romanes tells of behaved in the same way with a soap bubble rolling on the carpet He was highly interested, but could not make up his mind whether or not the thing was living, but after some hesitation he overcame his misgivings, approached cautiously, and touched the soap bubble with his paw. "The bubble, of course, burst at once, and I never saw astonishment more unmistakably expressed." †

Eimer gives an instance of the curiosity of cows: "As soon as I had my easel and sketchbook arranged the cows grazing about drew nearer and nearer, and stood in a circle around me, motionless and with necks outstretched, gazing at my paper as if to see what was going on. Finally, they came so near as to be annoying, and I was forced to drive them away with my stick. But again and again they renewed their attempt to penetrate the secret." ‡

Anschutz has portrayed the curiosity of horses in a very successful instantaneous photograph. As the photographer kneels on the ground busied with his camera, a number of loose horses surround him, pressing close and stretching their long necks inquiringly toward the strange objects. Scheitlin says of goats: "No

---

* Thierseelenkunde, ii, p. 342.

† Romanes, Mental Evolution in Animals, p. 157.

‡ G. H. Th. Eimer, Die Entstehung der Arten, 1888, i, p. 258.

single animal has more curiosity, unless it be the poodle. When a flock of goats is driven through a village, one and another will go into the houses, even into the rooms, and look about without concerning himself as to where the others are gone. He climbs over whatever he can, from mere curiosity, and sometimes goes to the second or third story of a house." * And the chamois is just as bad; they can be captured as can gazelles, by the display of a new or strange object, which so excites their curiosity that they forget the danger. Lloyd Morgan reports of his cat: "My cat was asleep on a chair and my little son began blowing a toy horn. The cat, without moving, mewed uneasily. I told my boy to continue blowing. The cat grew more uneasy, and at last got up, stretched herself, and turned toward the source of the discomfort. She stood looking at my boy for a minute as he blew. Then, curling herself up, went to sleep again, and no amount of blowing disturbed her further." † The animal had evidently accepted this new impression, and was satisfied to add it to her store of ideas.

A Fräulein Delaistre had a tame weasel, of which she says, among other things: "A notable quality of this animal is its curiosity. If I open a trunk or a drawer or look at a paper he must come and look too." ‡

The raccoon, too, is "curious to the last degree," says Weinland; of the one that has been described playing with a badger Beckmann writes: "One day he was too severe with the badger, which went off growl-

* Thierseelenkunde, ii, p. 207.
† Animal Life and Intelligence, p. 339.
‡ H. O. Lenz, Gemeinnützige Naturgeschichte, 1851, i, p. 164.

ing and rolled into his hole. After a time he put his head out on account of the heat and went to sleep thus intrenched. The mischievous 'coon saw that he could not expect much attention from his friend under these circumstances, and was about to set out for home when the badger suddenly awoke and stretched his narrow red mouth wide open. This so surprised our hero that he turned back to examine the rows of white teeth from every point of view. The badger continued immovable in the same position, and this excited the raccoon's curiosity to the highest pitch; at last he ventured to reach out and tap the badger's nose with his paw. In vain, there was no change. This behaviour of his comrade was inexplicable, his impatience increased with every moment, he must solve the riddle at any cost. He wandered restlessly about for a while, apparently undecided how best to pursue the investigation; but reaching a decision at last, he thrust his pointed snout deep in the badger's open jaws. The rest is not difficult to imagine. The jaws closed, the raccoon, caught in the trap, squirmed and floundered like a captive rat. After mighty scuffling and tugging he at length succeeded in tearing his bleeding snout from the cruel teeth of the badger and fled precipitately. This lesson lasted a long time, and after it whenever he went near the badger's kennel he involuntarily put his paw over his nose."

Mice and other rodents are curious,* and so are all kinds of seals. J. E. Tennent describes a hunt with tame buffaloes in Ceylon. If they are turned loose at night with lights fastened to their backs and bells hung

* See Hudson on the viscacha. The Naturalist in La Plata, p 293.

around their necks all sorts of wild animals, attracted by curiosity, come to look at them and are captured.*

That curiosity is a play closely connected with some of the primary instincts, such as flight and feeding, seems probable from the fact that it is found in some of the lower orders. Indeed, there are many facts in support of this view. Eimer tells us that the boys of Capri take advantage of the curiosity of lizards to catch these elusive creatures. "They make a slipknot in the pliable end of a long, slender straw and, lying down, hold the straw in front of a crevice where the lizard has just disappeared. Curiosity so torments the little creature that it comes nearer and nearer to examine the knot, until the boy seizes his chance to slip it over the head and secure his prize To excite their curiosity the boys sometimes make their noose of coloured membrane and wet the knot." † W. James says of young crocodiles that swarmed around him curiously, that they fled terrified at the slightest movement, but came back again directly.‡

Romanes, speaking of fish, says: "Curiosity is shown by the readiness, or even eagerness, with which fish will approach to examine any unfamiliar object. So much is that the case that fishermen, like hunters, sometimes trade upon this faculty:

> 'And the fisher, with his lamp
> And spear, about the low rocks damp
> Crept, and struck the fish which came
> To worship the delusive flame.' "#

---

* J. E. Tennent, Natural History of Ceylon, p. 56.
† Eimer, Die Entstehung der Arten, i, p 258.
‡ W. James, Principles of Psychology, ii, p. 429.
# Romanes, Animal Intelligence, p. 247.

It is a familiar fact that birds and fish and flying insects, as well as many mammals, are attracted by fire. J. S. Gardener noticed, while looking at an island waterfall, that one moth after another hurled itself into the cataract, probably attracted by the glittering water, as others are by flame.* The opinion of Romanes, that this is due to curiosity,† will hardly be controverted.

Turning now to birds, we may characterize them *en masse* as curious, so much so, indeed, that many of them fall victims to their curiosity, for all over the world hunters lure them by means of unfamiliar objects, which they approach to investigate. On islands uninhabited by man they will come up to the first human being they see without fear, the better to observe him. The crow family in particular possess this quality in excess; if a cane handle or almost anything is held near a caged raven he will come near it and examine it carefully from every possible point of view. Their efforts to get possession of and hide everything that comes in their way are further manifestations of curiosity. Parrots, too, behave in a similar way. Haast says that the curiosity of the keanestor impels them to examine everything that comes in their way. On one of his expeditions in the mountains he had with great difficulty collected a bundle of rare Alpine plants and laid them for a moment on a projecting rock. During his short absence a keanestor examined the collection and manifested his interest in botanical studies by pushing the whole bundle off the rock, never to be recovered. With ravens, as well as

---

* Nature, vol. xxv, p. 436.

† Mental Evolution in Animals, p. 279.

parrots, mental experimentation is connected with the physical, especially where the destructive instinct is concerned. Paske gives in the Feathered World (1881) an interesting description of a raven that he brought up. It delighted to fly into strange windows and do all sorts of mischief. He once entered, in this way, a room in the opposite house, found a collection of curios that had been left out of their case, and destroyed most of them. He showed his interest in the boys' ball games by stealing and hiding the ball. The following performance of his might have inspired Dickens to a special chapter in Barnaby Rudge: "One day he entered, through the window, a room where a military trial was being conducted, perching on the desk littered with writing materials and important papers, and refusing to be dislodged. He threatened with his bill every one who approached him, until I was sent for and carried him off."

If any strange object is held in a canary's cage he will examine it with great interest, turning his head first to one side and then the other, and it is most amusing to see the little creature crane his neck to look down at something under his cage, while he keeps up a succession of questioning peeps. Rey had a Carolina parrot which was so tame that he allowed it to fly about at will, much to the wonder and excitement of the domestic birds when this foreigner appeared among them. A sparrow was "so fascinated by the gay stranger that he followed the parrot about for a long time, sitting near it and gazing till the parrot returned to the window, without appearing to notice that I stood with a friend at the open casement."

The starling, the robin, the nightingale, the siskin, and many other birds have a great deal of curiosity.*

Last, I may mention the vulture, which is noticeable for this quality when young and will come near any one who displays a new and attractive object. Brehm's brother, in Spain, placed an owl in the vultures' cage, and describes the curiosity with which the occupants examined the newcomer. One young vulture approached the bird of night as it sat sulking in a corner, looked him over and began an examination of his feathers, an impertinence to which the owl responded with a sharp blow from his claw.

In most of these examples the animal is represented as seeing a new object and trying to find out what sort of thing it is—curiosity is expressed by approaching it, looking it over, tasting, etc. All this leads us again to the question whether animals may not have a kind of æsthetic perception. The case is quite different with these successive impressions from that of the coloured feathers and stones which they collect, for it is an established fact, as has been said, that among animals motion is more provocative of attention than anything else. Further, it is evident that imitative impulse is more easily awakened by movement than by any attribute of a body at rest. Accordingly, if that "inner imitation" that characterizes æsthetic perception can appear anywhere in animal life it may be looked for as a consequence of the observation of the motions of other animals, preferably individuals of the same species. Under the heading of imitative play it was shown that such movements do produce external imitation; so it

* Naturgeschichte der Vögel Deutschlands, iv, p. 16; ii, pp. 197, 203.

would appear that the animal, though aware of the stimulation to external imitation from optical and acoustic impressions, is able to hold it in check so that an internal excitation alone is produced by the imitative impulse, whose reflex in consciousness consists of "feelings of imitation." In order to illustrate my conception of the origin of such æsthetic feeling I venture to cite a progressive series of examples from human life. A boy on the streets sees some other boys chasing a comrade in play; he looks on for a few seconds, his interest constantly increasing, until he joins the pursuers. These few seconds of observation I regard as the primary form of æsthetic perception directed toward the movements that incite his impulse of imitation, for there is an inner imitation as an antecedent or point of departure for the outer. A boy takes part in a game involving complicated movements. He is taken prisoner by the opposing party and must stand in a base until one of his own side frees him. Æsthetic perception is manifest in the absorbed attention with which he enters into all the movements of his companions, for, while his impulse to external imitation is so far arrested by the laws of the game that it can not attain its object at once, this result follows as soon as the boy is at liberty to move from the base.—Suppose some witnesses of a race. Here the impulse to active imitation does not tend to external discharge. No one tries to leave his seat, but contents himself with expressing the feelings produced by internal imitation of the varying operations. Here we have the simplest and most primary form of pure æsthetic perception.—We are sitting in the theatre, the simulated actions and tones of voice are only the means of appealing to our sympathy and placing us mentally in touch with what is being played on the stage, and yet our facial expression corresponds in a certain degree with that of the actor.—Or suppose we are merely listening to a recital, we still feel all the sympathetic passion that words can produce. Indeed, the mere reading of a narrative is sufficient to produce that internal effect of imitation which consists in æsthetic pleasure. Don Quixote shows us how strong this impulse may be when he tries to realize the ideal which he has formed by reading. It is illustrated, too, by boys who read of a seaman's life till they can not be restrained from adopting his calling with all its hardships and dangers; by the suicides that have resulted from reading The Sorrows of Werther; and by the mystical religious life of saints, and the stigmata produced by auto-suggestion in many ecstatic fanatics. All these are externalized effects of æsthetic emotion.

A glance over these illustrations shows at once that those effects depending on the power of speech can not, of course, be attributed to animals, but that the cases of the boys at play are probably equally well applicable to them. All consciously imitative play must be preceded by that primary form of æsthetic perception which we have called "inner imitation," as, for example, when the monkey mimics his master, or when the starling, with head on one side, listens attentively to an air whistled in his presence. But, on the other hand, there are plenty of examples of attentive watching and listening without any external imitation. Most conspicuous in this class is the hearkening of the female bird to the song of the male. It can not be questioned that she experiences an internal sympathy with his excitement, for sometimes this feeling is so strong as to require some kind of outward expression, and she joins, though imperfectly, in the song of the male, and sometimes even takes part in his battles. A description already quoted says: "Sometimes a female is found on the arena taking up a position like that of the males and running about with them; but she does not long mingle in the strife, and soon runs away." No clearer proof could be desired that the female feels a secret sympathy in the love-plays carried on before her, for in such a case it evidently clamours for expression until the impulse to join in the song or dance is irresistible, as in the orgies described by Middendorf. Many birds arrange a regular stage or arena. Hudson says: "There are human dances in which only one person performs at a time, the rest of the company looking on, and some birds in widely separated genera have dances of this kind. A striking example is the rupicola, or cock-of-the-rock, of tropical South America. A mossy level spot of earth surrounded by bushes is selected for a dancing place, and kept well cleared of sticks and stones; round this area the birds assemble, when a cockbird, with vivid orange-scarlet crest and plumage, steps into it, and with spreading wings and tail begins a series of movements as if dancing a minuet; finally, carried away with excitement, he leaps and gyrates in the most astonishing manner, until, becoming exhausted, he retires, and another bird takes his place." *

There are examples on record, too, that seem to indicate that some of the higher animals observe the movements of others than their own kind with a sort of æsthetic perception. The most familiar of these is that of a dog looking out of a window. Schopenhauer considered this critical watching of passersby that can

* The Naturalist in La Plata, p. 261.

have no other intent than that of taking note of the various figures on the street as the most human quality displayed by animals. It is certainly comical to see a big dog with his forepaws on the window-sill gazing, for it may be half an hour, just as a man would do, with thoughtful, wrinkled brow, into the street.

But other animals, too, do the same sort of thing. A forester in Würtemberg had a tame doe, of which, among other things, he relates the following: "She likes to stand on the window-sill and watch what is passing outside." *

Among monkeys the *Cerocebus albigena*, a rather large black African ape, may be instanced. Pechuël-Loesche has described it in detail: "But he was drollest when some new problem exercised his busy brain, as when we used the astronomical instruments before him or carried on some unusual operation. He would sit on the ground or a trunk or barrel in the attitude of a deeply reflecting man, one hand holding his chin up and a finger pressed on his lips, while he followed our every movement, softly humming or grunting, and occasionally indulging in one of the philippics already described." (This species has a very loud characteristic roar.) † A. Gunzel contributes this about a tame magpie: "At the time of the morning recess he repaired to the playground of the school children, generally that of the boys, to watch their romps. At these times he expressed his satisfaction by jumping about and snapping his beak." ‡

The following story is told of a goose: "Some years

---

* Diezels, Niederjagd, p. 145.

† Loango Expedition, iii, p. 243.

‡ Die gefiederte Welt, 1887. See K. Russ, Allerlei sprechendes gefiedertes Volk, p. 74.

ago a goose excited considerable attention in a small town by its strange actions. Whenever the parish clerk came from the market with his great bell, as was the custom, to make a proclamation, a black and white goose left the flocks assembled at the brook and waddled hastily toward the circle of listening peasants. There she stood immovable all through the ceremony, with head outstretched as though she would parody the attentive attitude of the other auditors, until the bell was taken up. At this moment she set out to follow the officer to the next street, where she again took the listening attitude, and in this way accompanied the man all over the widespread town, only seeking her companions at the brook when he returned to his office. This habit was kept up for many months." * The famous parrot belonging to Director Kastner, in Vienna, always noticed when a bottle was about to be uncorked, and imitated the pop before it came, showing absorbed attention and anticipation.†

Two points of psychological interest are still to be noted. When I spoke of æsthetic attention, I did not mean to imply that æsthetic pleasure consists in conscious acts of attention, the word being used in the ordinary sense. If, for example, the female bird witnessing the performance of the males once attained to apperception, no doubt the imitative impulse would be roused just as with ourselves, without the conscious effort of attention. The question whether there may not be a constant unconscious anticipation may be answered affirmatively on various grounds, but this is not the place to explain them. Secondly, it may be re-

---

* Der zoologische Garten, vii. p. 238.

† K. Russ. Die sprechenden Papageien, 1887, p. 29.

marked that while the foregoing examples, of whose æsthetic character I have no doubt whatever, should be looked upon as only elementary expressions of æsthetic pleasure, they yet serve to show that the sphere of æsthetics is infinitely wider than that of the beautiful.

CHAPTER IV.

THE PLAY OF ANIMALS (*continued*).

Love Plays.

THE treatment of this class of plays in a separate chapter is justified not only on the ground of its importance to animal psychology, but also for two reasons inherent in its nature. The first of these reasons is that it embraces the vexed question of sexual selection, and the second reason is that this kind of play differs from all that we have previously considered in being, strictly speaking, not mere practice preparatory to the exercise of an instinct, but rather its actual working. Yet it is universally spoken of as play, and consequently our first question is, How far is this designation correct?

In considering it we are at once brought face to face with the problem of sexual selection, for Darwin regards all phenomena connected with love play as the direct result of the operations of this, his second great principle of evolution. Sexual selection, then, involves two distinct phenomena: on the one hand the conflict between males for the possession of a female, and on the other hand the preference of the latter for certain qualities or capacities in the former. Each of these phenomena is supposed to produce its own effect in the modification of characters. The first, being only a special case of natural selection, is challenged by no one. The selective principle involved in the second is not the mechanical law of survival of the fittest, but rather the will of a living, feeling being capable of making a choice, and is much like that employed in artificial breeding. Spencer has spoken of natural selection as a "survival of the fittest," and a fitting designation of this theory of sexual selection would be "a multiplication of the most pleasing."

Let us take an example. The male cicada has on one wing a vein set with fine teeth, on which he fiddles with the other wing. Only males can produce this music. "The ancient Greeks knew this, for Anacreon congratulated the cicadas, in a poem that has come down to us, because they had dumb wives." "Here is the key to the riddle. The origin of the musical apparatus is easily explained by means of the male's rivalry. If we assume that the females enjoy the music—and it has been proved that they do—then we see why and how a singing instrument was gradually developed from the male's wings and has been improved to its present perfection, for the female would always prefer the male that sang best. Thus the superior musical apparatus of the father would be inherited by his sons, and so on. In this way there must necessarily be much progress in the development of this function in the course of several generations, the preference of better singers constantly tending to improve the singing apparatus until it can be improved no further." * In the same way the musical performances of birds, the arts of flying and dancing, the strange and beautiful colours

* A. Weismann, Gedanken über Musik bei Thieren und beim Menschen, Deutsche Rundschau, lxi (1889), p. 51.

and forms, are all to be considered as "wedding garments," so to speak.

But many voices worthy of attention have been raised against this theory of a choice of the most pleasing by the female. Wallace takes the lead in this opposition, and many scientists agree with him either wholly or in part. I may mention Tylor,* Spencer,† Wallaschek,‡ Hudson,# Lloyd Morgan.||

Wallace has expressed his view in various of his works, the most important being the Natural Selection and the Darwinism, that Darwin's assumption of a kind of æsthetic taste in the female governing her choice is as far from the truth as is the assumption that the bee is a good mathematician. But more than that, he maintains that it is by no means certain that the female makes any choice at all. "Any one who reads these most interesting chapters (in Darwin's Descent of Man) will admit that the fact of the display is demonstrated, and it may also be admitted as highly probable that the female is pleased or excited by the display. But it by no means follows that slight differences in the shape, pattern, or colours of the ornamental plumes are what lead a female to give the preference to one male over another; still less that all the females of a species, or the great majority of them, over a wide area of country and for many successive generations, prefer exactly the same modifications of colour or ornament."^

---

* Alfred Tylor, Coloration of Plants and Animals, London, 1886.

† The Origin of Music, Mind, xv (1890).

‡ On the Origin of Music, Mind, xvi (1891).

\# The Naturalist in La Plata, chap. xix.

|| Lloyd Morgan, Animal Life and Intelligence, p. 407.

^ A R. Wallace, Darwinism, p. 285.

But we ask, What then is the cause of these phenomena, if there is no choice by the female? How do the beautiful colours and characteristic forms of male birds arise? Wallace answers these questions as follows: In the first place, it is not at all strange that animals should have colour. In all Nature colour is the rule, black and white are exceptions.* "The presence of some colour, or even of many brilliant colours, in animals and plants would require no other explanation than does that of the sky or the ocean, of the ruby or the emerald—that is, it would require a purely physical explanation only." † The kind of colours, however, is principally determined by natural selection. Colouring for offence and defence is very important in the animal world, a principle which was clearly recognised before the time of this misleading idea of sexual selection. Other peculiarities, such as broad white bands and white or coloured spots,‡ serve as distinguishing marks to those that live in companies.#

These marks are important not only in times of danger, when they make it easier for the young to follow the old ones, but they also form a kind of bond for the social life, and in addition to that probably serve a useful purpose in hindering the cross-breeding of closely related species. The symmetrical marking which renders the individual recognisable from either side seems to be for this purpose, as we conclude from the facility with which it is lost in domestication. To the same origin may be attributed the characteristic call

* Nat. Selection and Trop. Nature, 1891, p. 359.

† Darwinism, p 189.

‡ Many such marks are only visible while the animal is in motion, because they would expose it to danger when at rest (ibid.).

Like the tribal marks of savages.

of the male and the female's answering cry. "These are evidently a valuable addition to the means of recognition of the two sexes, and are a further indication that the pairing season has arrived; and the production, intensification, and differentiation of these sounds and odours are clearly within the power of natural selection. The same remark will apply to the peculiar calls of birds, and even to the singing of the males. These may well have originated merely as a means of recognition between the two sexes of a species and as an invitation from the male to the female bird. When the individuals of a species are widely scattered, such a call must be of great importance in enabling pairing to take place as early as possible, and thus the clearness, loudness, and individuality of the song becomes a useful character, and therefore the subject of natural selection." * Thus sexual selection would be absorbed in natural selection, and Wallace advances two principles to assist in the absorption Many characteristic markings and decorative colourings are, according to A. Tylor, closely connected with anatomical structure. Since the clearest colours show where the most important nerves run, their intersections form all sorts of figures. And "as the nerves everywhere follow the muscles, and these are attached to the various bones, we see how it happens that the tracts in which distinct developments of colour appear should so often be marked out by the chief divisions of the bony structure in vertebrates, and by the segments in the annulosa." †

* Darwinism, p. 284. We see here how Wallace came to change his mind about instinct.

† Ibid., p. 290. Tylor, for instance, finds in the zebra's stripes a picture of the spine and ribs. But why, then, is the symmetry so soon lost in domestication?

If, then, colouring is connected with nerve distribution it must be largely dependent on good health, and brilliant colour becomes an indication of robust health. This is true also of other kinds of external ornamentation, especially of the size of the tail. The perfect adaptation of animals to their environment produces in them a superabundance of vigour which contributes to the size and brilliance of plumage that we admire in such birds as the pheasant, parrot, humming bird, etc. To the question why this is the case with males alone it may be answered that the female has the greater need of protection. This view is supported by the fact that in general, female birds of those species that have well-protected nests are as brightly coloured as the males.

Wallace applies this principle to skill in flight and dancing as well as to ornamentation, the same principle of superabundant energy which we found in the Schiller-Spencer theory of play. "The display of these plumes will result from the same causes which led to their production. Just in proportion as the feathers themselves increased in length and abundance, the skin muscles which serve to elevate them would increase also; and the nervous development, as well as the supply of blood to these being at a maximum, the erection of the plumes would become a habit at all periods of nervous or sexual excitement." . . . "During excitement and when the organism develops superabundant energy, many animals find it pleasurable to exercise their various muscles, often in fantastic ways, as seen in the gambols of kittens, lambs, and other young animals. But at the time of pairing male birds are in a state of the most perfect development, and possess an enormous store of vitality; and under the excitement of the sexual passion they perform strange antics or rapid flights, as much probably from the internal impulse to motion and exertion as with any desire to please their mates." *

The act of singing, too, which was originally a means of recognition, "is evidently a pleasurable one, and it probably serves as an outlet for superabundant nervous energy and excitement, just as dancing, singing, and field sports do with us." †

These are the essentials of Wallace's theory. Selection through the female is excluded; at the most he thinks we may say that she prefers the "most vigourous, defiant, and mettlesome male," and so indirectly favours the ornamentation which results from abundant energy.

In this presentation of Wallace's theory I have maintained a careful distinction which is not made clear in his own works, but without which it is difficult, in my opinion, to understand thoroughly the meaning of his ideas I mean the distinction between the biological principles that would refer our problem to the familiar operations of natural selection, and such physiological theories as those of Tylor and Spencer. The former are of the greatest value, and will lead, I believe, to important modifications of the Darwinian system, while in the latter there is no inherent vitality, though Wallace seems to lay great stress on them.

Turning now to the secondary aspect of this theory, we set out from the fact that the characteristic marks and appendages of animals are closely connected with their anatomical structure, just as, in a common disease, an eruption occurs on the forehead which corresponds exactly to the distribution of the ophthalmic division of

---

* Ibid., pp. 294 and 287. † Ibid., p. 284.

the fifth cranial nerve. Supposing this to be a fact, still nothing has been said that is prejudicial to the theory of sexual selection—it must necessarily have some sort of physiological basis. However, I for one can not quite conceive how such developments as, for instance, a peacock's tail, can be derived from beginnings so insignificant, simply by a superabundance of energy. This is very delicate ground, for the hypothesis of surplus energy continuing through thousands of generations seems to me to accord little with the laws of natural selection, which are like the old laws of reward: they give with niggardly hand what is essential for the preservation of the species and no more.*

However, Wallace thinks that such extraordinary developments occur only when the species has acquired an assured position in life—in fact, "perfect success in the struggle for existence. . . . The enormously lengthened plumes of the bird of paradise and of the peacock are rather injurious than beneficial in the bird's ordinary life. The fact that they have been developed to so great an extent in a few species is an indication of such perfect adaptation to the conditions of existence, that there is in the adult male at all events a surplus of strength, vitality, and growth power, which is able to expend itself in this way without injury." †

But it is a well-known fact and a legitimate deduction from the principle of selection that such perfect adaptation to surrounding conditions produces a fixed type and precludes further development, just in proportion to its perfection. Thus, even if we suppose that

* The occasional surplus of energy arising from alternate waste and reintegration is, of course, quite another thing.

† Ibid., pp. 292, 293.

the ancestors of the peacocks, from the time when they attained a certain assurance of existence, were constantly in possession of surplus energy that favoured the production of strong (and useless) feathers, it is yet inexplicable how a still further development was attained, such as Wallace indicates. That such hindrances should arise before adaptation is out of the question, and it seems hardly possible after it, without the aid of sexual selection, for we see that success attained in the struggle for life prevents Nature from further directing the growing energies. The contest of males in which the strongest have the advantage would then come prominent forward as the only possible explanation. Wallace, however, has but cursorily referred to this principle and rightly, as I believe, for it is difficult to see how selection acting through the contests of courtship could directly favor the development of such peculiarities, since we can hardly suppose that surplus energy would find its only expression in them.

Perhaps Wallace recognised this difficulty when he wrote, "As all the evidence goes to show that, so far as female birds exercise any choice, it is of the 'most vigorous, defiant, and mettlesome' males, this form of sexual selection will act in the same direction (as natural selection), and help to carry on the process of plume development to its culmination." *

With these words, however hypothetical their form, Wallace overturns his whole argument, for if it is once admitted that the female chooses the strongest male, the chief point of the Darwinian theory is conceded. Whether her preference is for strength and courage

* Ibid., p. 293.

or for beauty is of little consequence; the important thing is that a choice is made.

Wallace's further deductions from the arts of dancing, flying, and singing will not detain us long. It is pretty well established that bird songs are inherited, generally speaking, and it seems quite as certain, if not more so, that characteristic dances and skill in flight have the same origin. Hudson says: "But every species or group of species has its own inherited form or style of performance; and however rude or irregular this may be, . . . that is the form in which the feeling will always be expressed." *

If this is true, mere surplus energy in the individual can not explain it. Of course the Lamarckian theory has no trouble with it. Its advocates can say with Hudson, "If all men had agreed at some period of race history to express the joyful excitement which now has such varied manifestation, by dancing a minuet, and if this dance had finally become instinctive, men would be in the same case that animals are in now." † But Wallace is very sceptical about the inheritance of acquired characters, and takes special pains to refer instinct finally to natural selection.‡ Whoever agrees with him in this must cast aside his Spencerian theory of courtship, for it stands or falls with the Lamarckian principle. Once grant that there is no inheritance of individually acquired habits, and that choice by the female is not influential, then these phenomena, which are of too great importance to the species to be dismissed as a mere discharge of surplus energy, however favour-

* The Naturalist in La Plata, p. 281.

† This example is well calculated to show how improbable the inheritance of acquired characters is.

‡ Darwinism, p. 441.

able that condition may be for them, must be referred at once to natural selection.*

The case is quite different, as I have remarked, with the first element of Wallace's theory. Here the gifted author advances his original ideas and reaches conclusions which are calculated, in my opinion, to seriously modify the Darwinian theory of sexual selection. Taking, as an example, the parrot which is commonly of a green ground colour with stripes of yellow, red, and blue, Wallace would say that through adaptation to life in the woods the green colour serves as a defence, while the stripes are distinguishing marks for purposes of recognition, and we have brilliant plumage explained satisfactorily without any reference to sexual selection, which can not, then, have the range that Darwin attributes to it in accounting for the colouring and other ornamentation of animals.

Quite as convincing, too, is the argument against the exercise of æsthetic judgment, comparison, and selection in pairing. I am even inclined to go further than Wallace and exclude the conscious choice of even the strongest and bravest, which he seems disposed to admit, but I do not on that account imagine that the Darwinian hypothesis is refuted.

Going on to consider bird-songs, Wallace says: "The peculiar calls of birds, and even the singing of the males, may very well have originated merely as a

* [It is precisely at such critical junctures as this that the principle of Organic Selection (see above, p. 64, and the Appendix) is needed to relieve the strain on natural selection If there be any preferential mating—even the little conscious choice admitted by Wallace, or the more physiological sort suggested by Groos—it would set the direction in which natural selection would accumulate variations.—J. M. B.]

means of recognition between the two sexes of a species, and as an invitation from the male to the female bird." These acoustic signals become very important when the members of a species live far apart, and are of especial service to migratory birds whose males first arrive at the destination and call to their mates to follow. The male that is distinguished for the "clearness, loudness, and individuality" of his song would succeed first in accomplishing this, and has thus an advantage that may be decisive in the struggle for life. But in that case the clearness, loudness, and individuality of his song would be a sufficient object for the operation of natural selection.* A close examination of these citations shows, I think, that while they modify the Darwinian theory very considerably, they do not exclude it. It would, indeed, be absurd to affirm that all bird-songs originate in a conscious æsthetic and critical act of judgment on the part of the female. A conscious choice either of the most beautiful or the loudest singer is certainly not the rule, and probably never occurs at all. But is it not still a choice, though unconscious, when the female turns to the singer whose voice, whether from strength or modulation, proves most attractive? Even if the song is primarily a means of recognition or an invitation from the male, still the psychological effect must be that the female follows the songster that excites her most, and so exerts a kind of unconscious selection.

But this is essentially the Darwinian idea, since, though there is indeed no conscious æsthetic selection,

* On this point I agree with E von Hartmann's penetrating criticism of the Darwinian theory. See especially his Philosophie des Unbewussten, iii, p. 435.

a kind of unconscious choosing does take place which is in a peculiar sense sexual selection, for the female is undoubtedly more easily won by the male that most strongly excites her sexual instinct. That such a selection as this is difficult or even incapable of proof from its very nature is no argument against its existence.

Wallace says, indeed, that all the things a young man may do to make himself acceptable in the eyes of his beloved, while they do perhaps please her, have no influence in inducing her to accede to his wishes.* But is this a fact? A conscious influence would scarcely be allowed to them, but will not a fine figure, good address, noble carriage, and even tasteful dress prove a powerful spur to unconscious choice? Will not the soldier in his handsome uniform be more acceptable than the same man in his working blouse? Has not the actor or the singer who has distinguished himself a better reception than a man quite his equal, but engaged in a commonplace business?

And now, putting aside everything that distinguishes man from the other animals, all our appreciation of intelligence and culture, all higher æsthetic influences, all considerations of a practical and material nature—conceive such a human race and suppose a condition of absolute free love with every spiritual ground for preference removed—must we not suppose that such (impossible) human beings left to the mere processes of evolution would become stronger and more beautiful in the course of a hundred or more generations?

I can not, then, admit that sexual selection is entirely subverted by Wallace's conclusions. If we accept his

---

* Darwinism, p. 286.

theory of bird-song, an involuntary selection based on the strongest sexual excitement takes the place of a choice of the most pleasing, and we may assume the like in regard to the other arts of courtship. Having conceded so much, we must also admit that the excitement may be augmented by the display of unusual colours and forms, making sexual selection influential upon these as well, but that it was prepared for by the factors introduced by Wallace, in a much more extended way than was understood by Darwin.

The Darwinian principle thus improved seems to me to be the only one in existence that has the least value as a working hypothesis. It can not, of course, be said to be as well established as is the principle of natural selection, but it is materially strengthened by the substitution of involuntary yielding to the strongest impulse for conscious æsthetic choice on the part of the female.* It must be borne in mind, however, that such selection does not, as a rule, imply a direct rejection of less favoured suitors, but owes its chief effectiveness to the advantage it offers the favoured male in securing the earlier birth of his children.

As we are now on hypothetical ground, the following note may not be out of place in conclusion. Professor Ziegler, of Freiburg, says, in the course of a private communication, "Among all animals a highly excited condition of the nervous system is necessary for the act of pairing, and consequently we find an exciting playful prelude very generally indulged in." The germ of a still more far-reaching modification of the theory of sexual selection seems to me to be contained in this

---

* This idea has been more fully carried out by Von Hartmann (*loc. cit.*).

indisputable fact. In the first place, it is certain that in general before any important motor discharge there is apt to be a preparatory and gradually increasing excitement. A period of rapidly increasing irritation which causes various reflex movements generally precedes a wrathful onslaught, as angry dogs illustrate no less than Homeric heroes. While we find this introductory stage, which is easy of explanation physiologically, reduced to a minimum in the instinct for flight and in the spring upon prey, it appears to be at the other extreme in the courtship of many animals, for we find a long-continued preliminary excitation necessary, which presents strange peculiarities. This fact seems to me to suggest very strongly the probability that in order to preserve the species the discharge of the sexual function must be rendered difficult, since the impulse to it is so powerful that without some such arrest it might easily become prejudicial to that end. This same strength of impulse is itself necessary to the preservation of the species; but, on the other hand, dams must be opposed to the impetuous stream, lest the impulse expend itself before it is made effectual, or the mothers of the race be robbed of their strength, to the detriment of their offspring.

If this be granted, all the rest follows easily enough. The most important factor in maintaining this necessary check is the coyness of the female; coquetry is the conflict between natural impulse and coyness, and the male's part is to overcome the latter. This is accomplished most easily by pursuit, and at last by what appears sometimes as violence, but probably is not really such, but only a necessary stage in the attainment of the requisite pitch of excitation. There are other means as well; for instance, scent in many animals, that is useful as a means of recognition, is a powerful agent here; contact, too, plays its part with most animals,* as well as the regular love plays, such as dancing, flying, and singing.

And since with these is connected the display of brilliant colours and striking forms, the intensifying of performances that were perhaps originally intended to serve other purposes may help to overcome the female's reluctance.

In all this we have attempted to indicate the outlines of a view which would so transform the original Darwinian principle that if fully carried out we should have to consider it a new theory. Sexual selection would then become a special case of natural selection. If the point of departure for this idea be granted—namely, that the excited condition necessary for pairing, and also a certain difficulty in its execution, are both useful for the preservation of the species—we find the whole series of phenomena related to the subject so much more simply and satisfactorily explained that no one, it seems to me, can hesitate to decide in favour of the hypothesis. Instead of a conscious or unconscious choice, of which we know nothing certain, we have the need of overcoming instinctive coyness in the female, a fact familiar enough, but hitherto not sufficiently accounted for. Then the question is no longer which among many males will be chosen by the female, but which one has the qualities that can overcome the reluctance of the female whom he woos. How great a difference this is will appear from the fact that in the well-founded opinion of the Mullers the choice itself, the betrothal of the birds, as it were, takes place before the breeding time. "Long before the springtime, with all its enticements

* *Cf.* Espinas, Les Sociétés animales.

to love, the young birds have chosen their mates, unseen by any but the closest observers. It is a common mistake to suppose that the marriage bond is first assumed in spring. Rather to this time belongs the male's first solicitations for his mate's consent to sexual union, and this has been falsely called pairing." * If this opinion, so emphatically expressed, is correct, the explanation of the phenomena of courtship by means of conscious or unconscious choice is irretrievably damaged. Our view, on the contrary, would be in perfect accord with it, whether there had been a previous choice or none at all. Reproduction would be assured to the male who possessed the qualities and capacities necessary to conquer the instinctive reluctance of the mate.

This explains, too, why the dallying of birds that have lived in wedlock for a long time is repeated year after year, and indeed much oftener, although there can certainly be no further selection by the female.† And, finally, our hypothesis applies equally well to plays by masses and whole flocks together, and to those cases where the female takes part in the flying and singing, which present great difficulty to the Darwinian theory, and yet it does not preclude the possibility of a conscious or unconscious choice in Darwin's sense.

Before going on to the second part of our subject, I wish to notice the common objection that what we call the arts of courtship are frequently practised at other times.

This objection is expressed most clearly by Spencer,

---

* A. and K. Müller, Thiere der Heimath, i, p. 4.

† The zoölogist H. E. Ziegler says, in a notice of this book: "This circumstance favours the author's view, for there could hardly be sexual selection in such cases, but rather excitation only."

Wallace, and Hudson. One of the passages already cited from Wallace will serve as an example. He says that when there is a surplus of energy the animal indulges in all sorts of strange motions and vocal exercises. This happens, it is true, most commonly at the mating time, when the animal is in full possession of all his powers, but may occur at any time when there is superabundant vigour.

I have already pointed out that such a conception of hereditary instinct must to have any value be supported by the theory of the inheritance of acquired habits. Even Darwin, who concedes the Lamarckian principle, has expressed himself in opposition to the view that such phenomena may be regarded as expressing a general state of exhilaration, with only a secondary application to courtship.* He quotes from an article † and from letters of Joh. von Fischer that a young mandril, when he saw himself for the first time in a mirror, turned round after a while with his red back toward the glass, just as many apes do when they see strangers looking at them. (Brehm quotes an ancient description of a mandril by Gesner: "This animal was brought to Augsburg with great wonder and exhibited there. On his feet he had fingers like a man's, and when any one looked at him he turned his back.") Other cases are recorded where the animal apparently desired to display what he considered his greatest beauties and attractions, just as these monkeys show to the observer their most highly coloured parts.

How shall we account for these facts? Can they be the effects of ordinary reflexes answering to any

* Nature, November, 1876

† Der zoologische Garten, April, 1876.

excitation? That is not likely with movements so significant. Or shall we say with Fischer that the reason monkeys enjoy being stroked or scratched on bare spots is because such sensations are associated with the friendly offices of comrades in removing thorns, etc., and from this their modes of greeting have developed? Is it not a thousand times more natural to suppose that such instincts are primarily for the purpose of sexual excitation, though they are sometimes connected with other stimuli? I do not think that any one can seriously doubt where the greater probability lies.

These remarks also apply to dancing, evolutions in flight, contortions of the body, erection of feathers, making strange noises, as well as the calling and singing of amorous animals. In all these we see instinctive acts performed for the purpose of arousing excitement, usually in both sexes. As the ape exhibits such actions most unmistakably, I have cited his case to avoid the possibility of misconstruction, for none can deny their connection with the sexual life, or attribute them to any ordinary excitement.

In order to arrive at a satisfactory position, however, on this question, we must occupy middle ground between the contestants. We must admit that in most cases the actual basis for the arts of courtship is to be found in general excitement reflexes, or even in those of quite a different origin. This basis consists partly in such reflex motions as result from any strong excitation, such as restless fluttering, running about, skipping, and trembling, and further in the reflexes that are commonly awakened in the face of an enemy, such as inflation, erection of hair or feathers, lifting the voice, etc. These are obviously the material from which Nature has derived the peculiar arts of courtship in all their variety, and these arts, as we have seen, are then extended to occasions which have no sexual meaning.

But what is our justification for calling this play? If the adult bird practises his skill in flight and song out of season and simply from good spirits, that indeed is play, and the gambols and dallyings of young immature animals are as much play as their romping is.

But, apart from these, it is common to speak of the arts of actual courtship in the same way, and this fact requires some explanation, though I confess I can not find one that is entirely adequate. The fact seems to be simply that the evolutions of birds on the wing, their songs and dances, and their naive display of what adornments they possess, impress us as playful, and we have fallen into the habit of speaking about all animals in the same way. But who knows that a mistaken analogy has not led us far astray? When a skater sees his beloved on the ice he displays all his skill before her, and a good dancer does the same at a ball; a man in love actually walks straighter and dresses better, and the power of song has its uses, too, in human courtship. When all this happens, we say the man is playing a part, is trying to appear stronger, more skilful, better looking, more sympathetic, etc., than he really is, and even if all the conditions of our definition of play are not fulfilled we must consider his conduct in that light But are we justified in extending the analogy to the animal world on such grounds as these? Certainly not, for apart from uncertainty of any far-reaching correspondence between human and animal life in their higher aspects, such a proceeding would involve the fallacy of comparing phenomena that are not even externally alike. To make them so, the amorous gentleman should lift his voice at the sight of his lady, should indulge in all sorts of capers and dances, and instead of the fingering of his beard merely it should rise on end of itself, etc. Since these things do not transpire, we are precluded from drawing any conclusion from human analogy as to the psychological significance of animal courtship.

In fact, there seems even to be a direct contrast. The youth desirous of showing off his good looks or skill in any art acts voluntarily, he consciously plays a part when that is necessary to his purpose. The animal, on the contrary, acts reflexly, following a blind propensity. His condition of excitation calls into activity certain motor tracks, and the animal obeys the impulse, unconscious that he is making a display of his attractions.

I have exaggerated this contrast purposely, to prove that such a crude dualistic conception of courtship as exhibited by men and animals is too much like the Cartesian view. But this statement of it requires modification, for, on the one hand, the young man playing the agreeable is not so entirely governed by reason as might appear, for blind propensity has as much to do with his actions as reflective choice has, and, on the other hand, the higher animals, and especially birds, exhibit such a degree of intelligence that I consider it nearer the truth to affirm than to deny a consciousness of self-exhibition in their displays of beauty and dexterity. There is, of course, between affirmation and denial the safe but fruitless position of the sceptic with his doubting shrug, but I think it is more honest to meet the question squarely and lay before the reader sufficient examples to justify an intelligent judgment upon the characteristics of a playful act.

The difficulty of definition is greater here than with other classes of play, for the reason that it is necessary to get at the psychic or inner features of the phenomenon. We established as a fundamental principle of our inquiry the fact that play is, to state it briefly, not exercise of, but practice preparatory to, instinctive activity. When, as in the case of young animals, the practice is obviously preparatory,* there is no occasion to speculate on the probable psychic accompaniments in order to establish its playful character. But here we are confronted with acts that are performed at the time for the actual exercise of the instinct, and consequently in their external manifestation appear as serious means to a real end. In such cases, therefore, only the psychic significance which I before put aside as a secondary consideration can decide as to the genuine playfulness of an act.

The bird performing his fantastic evolutions of flight and dancing before the object of his affections is not playing, so long as he only discharges the motor functions prescribed by heredity. Sexual excitement would produce the reactions necessary for courtship without anything taking place in the creature's mind other than takes place there when he involuntarily flies away at the sight of an enemy. All the complicated acts of courtship would then be nothing more than physiological results of excitation, a direct exercise of instinct for serious ends, not in any sense play. Familiar facts show that this may often be the case. It frequently happens that excitation unconnected with sex—such, for instance, as that produced by the sight of a foe—calls forth the manifestations usually associated with courtship, and not only those that might

* [The author's antithesis between Ausübung (exercise), on the one hand, and Einübung (practice) with Vorubung (preparation), on the other hand, can not be fully rendered by single English words.]

be calculated to inspire fear, such as erection of the feathers, etc., but even exhibitions of flight and singing. Canaries commonly trill shrilly while fighting, and Brehm says that the lapwing, which is very irritable during the nesting period, becomes wildly excited, sounds his mating call, and tumbles about in the air at the approach of a man or an animal. Since, then, the excitement of anger can produce these effects reflexly, it is probable that that of courtship may frequently act in the same way; indeed, among the lower orders this is probably the rule.

On the other hand, we are forced to remember that the acts of the higher animals are generally accompanied by quite complicated psychic processes. Taking birds again as our example, it must be conceded that an enlightened animal psychologist is obliged to attribute to them a highly developed intellectual and emotional life. "Adequately to treat of the intelligence of birds," says Romanes, "a separate volume would be required." * And we know from many independent observations that are mutually confirmatory to a remarkable degree that this statement is not exaggerated. Pigeons recognise a voice after many months' absence, and a bullfinch belonging to one of the Mullers did so after nearly a year.† Tame storks answer to a familiar name.‡ It is well known that birds dream, and parrots sometimes talk in their sleep; the emotions of love and sympathy are very active; the conjugal fidelity of many species speaks for the finer feelings, the wedded pair evincing the deepest grief on being

* Animal Intelligence, p. 266.
† A. and K. Müller, Thiere der Heimath, i, p. 114.
‡ Naumann, ix, p. 256.

separated and the greatest joy on reunion; crows, in spite of their shyness of a gun, hover about a comrade that has been shot instead of taking flight; parrots and storks revenge an injury after a long interval and rejoice diabolically over the success of a piece of mischief. The behaviour of peacocks and turkeys points to the conclusion that birds can be vain, but might be thought insufficient to prove it if it were not for the pride of talking birds over their accomplishments; Darwin has conclusively proved that they take pleasure in beautiful colours and musical sounds, many birds drop shell-covered prey from a height in order to break the shell; the teachableness of canaries, finches,* and other birds is astonishing, they learn the most difficult compositions; crows have been known to conduct trials, where by common consent some unpopular members of the community were condemned and executed; swallows, in whose nest a sparrow has established itself, wall up the entrance so that the intruder perishes.

This list might be extended indefinitely, but these few examples, which I have purposely chosen as some of the most remarkable among well-authenticated cases, clearly indicate mental endowments of a high order in the birds concerned.

This being established, it must be admitted that the ardent male who performs his flyings and dancings again and again before his mate and invariably succeeds by such methods in overcoming her reluctance, may well be quite conscious of what he is doing. That satisfaction in his ability to talk, which the parrot shows so plainly, and which appears so early in the child, is probably akin to the feeling which swells the breast

* See Naumann, v, p. 137.

of a bird when he conducts his courtship successfully. And thus it may come about that the actual exercise of instinct for a real end may, to a certain degree, have the psychological aspect of mere play Just as strong men sometimes undertake severe physical labour and derive pleasure from it—that pleasure in power which imparts a playful character to the most serious work—so we may suppose the wooing bird enjoys his own agility and skill, nor can we deny to him the satisfaction that makes their exercise a play.

Thus the dance of courtship may be considered psychologically as having the character of a movement play, though it is not actually play in itself considered. Further, it seems to follow, from the admission that the female enjoys witnessing such blandishments, that the male must be conscious of giving her that pleasure—that is, is conscious that he is making a display—and in this, too, the act which is a serious exercise of instinct takes on the psychological aspect of play. Thus the crude dualistic contrasting of human and animal courtships is shown to be unwarranted

Taking a general view of love plays, I distinguish five separate classes, as follows:

1. Love plays among young animals.
2. Courtship by means of the arts of movement.
3 Courtship by means of the display of unusual or beautiful colours and forms.
4. Courtship by means of noises and tones.
5. Coquetry in the female.

1. *Love Plays among Young Animals.*

Among animals that have a period of youth the sexual instinct usually finds expression in some sort of play long before maturity. This is especially noticeable in mammals, some of which even in infancy make efforts to produce the movements necessary for pairing, a fact which can only be explained as practice for later life. Such phenomena are common among young dogs and apes, and Dr. Seitz, in Frankfort, noticed them in an antelope only six weeks old. While there are cases, especially among monkeys, where there is so much excitement as to render the playful character doubtful, still, as a rule, it is attributed to youthful sportiveness. According to Dr. Seitz, it sometimes happens in these games that the sexes change their parts, the male coqueting and the female pursuing. Chr. L. Brehm has noticed this, too, in the case of the golden-crested wren.

Much detailed material has been collected by ornithologists to show that songs, dancing, and flying evolutions are extensively practised by young birds in their first autumn, too early to serve the purposes of reproduction. This is genuine play, practice for instinctive activity quite as much as are the chasing and fighting of young animals. "The song of birds," says the elder Brehm, "appears to be the expression of love, for it begins with many shortly before pairing and ceases altogether after it, and with those that sing all through the summer, as the field lark does, the pairing season lasts as long. Caged birds are no exception to this, for most of them lose their natural or hereditary song or never acquire it, as, for instance, the wood lark, the red linnet, and many others.

"Awakening love impels some birds in captivity to sing as usual, and they also breed in that state, but the majority lose their power to do the latter and sing only as the effect of rich food and *ennui*. But the most noteworthy thing about the whole subject is that their love is awakened long before breeding time, usually in the first autumn of their lives. This fact has not been announced before, and should be supported by weighty proof, therefore I will now proceed to name the birds whose young I have myself heard singing in the autumn. . . . Young magpies (*Corvus pica*) produce in September, often in August and October as well, the long metallic notes that characterize them in spring, just before pairing. . . . I have often heard the *Picus viridicanas* piping in September as beautifully as in April, and indeed the *Picus major* sometimes hums in the autumn, picking absently meanwhile among the dry branches just as he does in spring. The crossbill and some woodpeckers sing before they have shed their first feathers. Young house and field sparrows not only chatter and chirp, but swell up their throats and peck at one another just as they will do at the pairing time next spring. Red linnets begin their song while still in their baby clothes, learn it perfectly while moulting, and even in winter, if the weather is mild, join in singing with their elders. The wood lark sings as soon as his first moulting is past, not only while at rest, but mounting aloft as in spring, floating about as he sings. All the titmouse family sing, the swamp titmouse producing exactly the note that accompanies breeding, and in October, 1821, I saw one approach his mate with all the manifestations that precede pairing in the spring, while she dropped her wings and spread her tail." Brehm goes on to mention similar songs and actions on the part of starlings, water wagtails, willow wrens, black- and heath-cocks, and a great variety of other birds, and says in conclusion: "The fact that pairing does not follow these demonstrations proves their dissimilarity to those of domestic fowls. The young cock is physically developed very early and ready for pairing in the first autumn of his life, but, with the exception of the crossbill, this is not the case with the birds that have been mentioned. The awakening of love seems to fill these little creatures only with a beautiful tenderness, which inspires them to express their joy in song and other demonstrations.* Hudson tells us that many species of American woodpeckers engage in a kind of duet which is practised in their earliest youth. "On meeting, the male and female, standing close together and facing each other, utter their clear, ringing concert, one emitting loud, single, measured notes, while the notes of its fellow are rapid rhythmical triplets; their voices have a joyous character, and seem to accord, thus producing a kind of harmony. This manner of singing is perhaps most perfect in the oven bird (*Furnarius*), and it is very curious that the young birds, when only partly fledged, are constantly heard in the nest or oven apparently practising these duets in the intervals when the parents are absent; single, measured notes, triplets, and long, concluding trills are all repeated with wonderful fidelity, although these notes are in character utterly unlike the hunger cry, which is like that of other fledglings." †

It would seem, then, to be firmly established, among birds at least, that the arts of courtship are practised as youthful sport before the time for reproduction. In choosing the examples cited from Brehm I have intentionally included some that refer to flying and dancing motions, as well as to singing

---

* Brehm, Beiträge zur Vögelkunde, ii, pp. 747–756.

† The Naturalist in La Plata, p. 256.

2. Courtship by means of the Arts of Movement.

Beginning with mammals, some of the examples cited of fighting dogs belong here as properly. The amorous dog, too, in contrast to the feline tribe, which does not seem to have any special courtship movements, indulges in what might almost be called a dance. The motions are like those with which he approaches an enemy, especially the stiff-legged gait, the rigid tail, and the erect carriage of his head. The fact that a vain dog will behave in the same way when allowed to carry a cane proves the consciousness of self-exhibition.*

The stone marten proudly lifts his head at the approach of a female, his tail is curved, the limbs stiffened, the hair rises on his back, and his whole aspect suggests the utmost vigour.†

The fish otter tumbles and splashes around his chosen one in an extraordinary manner, during which performance his eel-like tail is in constant motion and the sinuous body is as often above as beneath the surface of the water.‡ The buck delights to follow a doe about until their breeding time in July or August, and, according to Diezel, the same thing is repeated in November, but this time without result.# The following interesting description of the action of some antelopes is by Schweinfurth: "About five hundred paces from the road we saw a group of sporting antelopes. Their manner of playing suggested a marching procession

* I have noticed, too, that at such times a fox terrier whirls around with rapid springs.

† Müller, Thiere der Heimath, i, p. 368.

‡ Ibid., p. 380, and Wohnungen, Leben und Eigenthümlichkeiten in der höheren Thierwelt, p. 204.

Diezel's Niederjagd, p. 142.

with an invisible leader. They followed one another in pairs, forming circles in the shaded wood as if they were in an arena. Other groups of three or four stood by as spectators, or from time to time joined the circle This went on until my dog disturbed and scattered the assemblage, but I had plenty of time to observe what I have attempted to describe. I believe that it was the breeding time of the animals, and for that reason they were oblivious of the approach of danger." Brehm says of the water rat: "Both sexes indulge in long-continued gambolling before they pair The male behaves very strangely. He turns so rapidly as to make a whirlpool in the water. His mate looks on with apparent indifference, but must secretly enjoy his exhibition, for usually when it is finished she receives him with favour." The whale in love "turns over on his back, stands on his head, lashing the waves with his tail, leaps up with his giant bride sportively above the water and performs other antics."

Observations on birds are exceedingly copious in this connection. Two kinds of motion can be distinguished among them, which though sometimes found in combination are quite unlike. I mean flying evolutions and dancing motions. Taking flight first, we have Brehm's description of the blue-throated warbler: "In sunny weather it tumbles about in the air and performs the strangest evolutions, plunging headlong downward it often turns a complete somersault, as Naumann * says. Then mounting slowly upward once more he flies like a dove, with quick movements of the wings and apparently with no object in view." Azara, describing a small finch which he aptly named Oscilador,

* Naturgeschichte der Vögel Deutschlands, ii, p. 164.

says that early and late in the day it mounts up vertically to a moderate height, then flies off to a distance of twenty yards, describing a perfect curve in its passage; turning, it flies back over the imaginary line it has traced, and so on, repeatedly, appearing like a pendulum swung in space by an invisible thread." * Audubon thus vividly portrays the American night hawk: "Their manner of flying is a good deal modified at the love season. The male employs the most wonderful evolutions to give expression to his feelings, conducting them with the greatest rapidity and agility in sight of his chosen mate, or to put to rout a rival. He often rises to a height of a hundred metres and more, and his cries become louder and more frequent as he mounts, then he plunges downward with a slanting direction, with wings half open, and so rapidly that it seems inevitable that he should be dashed in pieces on the ground. But at the right moment, sometimes when only a few inches from it, he spreads his wings and tail and turning soars upward once more." The same authority describes the mocking bird as fluttering about his mate and regularly dancing through the air. The whitethroat leaves his perch in a tree top while singing, rises ten to twenty yards and lets himself fall, still singing, either fluttering in a slanting direction or, with folded wings, almost perpendicularly.

The reed bird, while his mate is sitting on the nest, flies up in the air diagonally and floats with his wings held so that they nearly touch over him. The wood lark mounts in the same way, constantly singing, and after describing one or more circles falls or plunges down and slowly returns to the tree from which he set out.

* The Naturalist in La Plata, p. 263.

The siskin, crossbill, many kinds of pigeons, the lapwing, golden plover, and various other birds behave in a similar way at their breeding time. I close this part of the subject with Naumann's description of the snipe. In the pairing time "the male flashes like lightning from his place in the marsh, first on a slant and then winding upward in a great spiral to the sky. He goes so high that even on bright days only the strongest eyes can follow him. At this great height he floats about in circles and then shoots down perpendicularly to the ground with wings widespread and motionless." *

The marked similarity in the evolutions of such various birds must have attracted the attention of any reader of this collection of examples, which might be enlarged indefinitely by the addition of numberless others of a like character. Especially noticeable is the practice of that bold flight upward and then the rapid or slow return; it is peculiar and yet so common that its explanation seems a riddle difficult to solve. May there not be something in the fact that such a movement shows the under side of the bird's body to his mate? The kite, however, is said to take her with him on his flight, and in that case shows more of the upper part of his body. Yet once granted the operation of the instinct, and we may easily assume that the bird's gliding downward through the air is a delightful movement play which must be about as much like our coasting on snow as travelling on rubber tires is like the jolting of a dray wagon.

Among storks and preying birds the female generally participates in these flights. "It is a noble

* Naumann, *loc. cit.*, viii, p. 327.

sight," says Naumann,* "and has a quality of stateliness when a pair of storks in fine weather and at the beginning of their pairing time, for then they seem to enjoy it most, circle up in the air higher and ever higher, and at the top of the gigantic spiral disappear in the clouds." Whole flocks of cranes make these circles together, when the weather is fine and they are not hurried. Falcons and ravens rise in pairs to a great height and describe noble curves. Crown Prince Rudolf, of Austria, thus describes the kite: "In the spring, at pairing time, some idea of their powers of flight can be formed. Exhilarated by the knowledge of their love, the pair mount high in the air and move in circles. Suddenly one or the other drops, with wings relaxed, almost to the water, skims along rapidly in broken lines for a short distance, then turns and hastens upward once more, shakes like the kestrel, and performs some wonderful evolutions." Naumann says, referring to the buzzard: "It is a treat to watch their gambols above their nest in fine weather, how the pair slowly circle upward without moving their wings, the male gradually outstripping his mate. He then lets himself descend from a great height with a peculiar vibratory motion of the wings, repeating this performance over and over for perhaps a quarter of an hour."

The other kind of movement play common among wooing birds is the dance performed either on the ground or among the branches of trees. If skill in flight serves to display the male's beauty and agility to his mate, dancing is better calculated to call attention to and emphasize brilliant colours and advantages of figure.

* Naumann, ix, pp 250, 361.

I will cite only a few cases, selecting those in which the motions seem to me of an unmistakably exciting nature.*

First I may notice the crane, which is one of the most intelligent of birds, for in its actions we can see clearly how genuine courtship may become playful.

To corroborate my statement about the intelligence of cranes I give this description: "Herr von Seiffertitz had a crane that he captured when young and downy. He was allowed the freedom of the premises, and when he was a year old followed his master for long walks, separated quarrelling animals, went to pasture with the herds, drove in young cattle that strayed, turned away beggars, and quieted restive horses. When he was hungry he went to the window and called, and if his water was not fresh he threw it out and called for more. He had a special liking for the bull, visited him in his stall, kept flies off him, answered when he lowed, and accompanied him to the meadow, dancing about him at a prudent distance, and stopping now and then to make ridiculous bows. If his master scolded him, the crane stood in the most dejected attitude, with his head bowed down to the ground" † That an animal of such intelligence should dance purely to amuse himself is not at all surprising Brehm says: "The crane delights, when in the mood for it, in vigorous leaps, excited

* A third kind of movement play would be the skilled swimming of aquatic birds, but of this I know but one example in literature Female wild ducks, just before pairing, swim around their mates, nodding their heads and quacking loudly (Naumann, Naturgeschichte der Vögel Deutschlands, xi, p. 600). I have also seen a pair of swans sporting. They would dip their heads deep down in the water together, and when they drew them up the neck of one would often be lying across that of the other.

† Lenz, Gemeinnützige Naturgeschichte, ii, p. 312.

gesturing, and strange positions. He twists his neck, spreads his wings, and regularly dances; sometimes he stoops repeatedly in rapid succession, spreads his wings, and runs swiftly back and forth, expressing in every possible way unbounded joyousness, but through it all he is always graceful, always beautiful." * "The peacock crane stands on a sand bank and begins to dance at the slightest provocation, sometimes nothing more than the fact that he has stepped on a hillock. The dancer often springs as high as a metre from the ground, spreads his wings and sets his feet down mincingly. I do not know whether both sexes dance, but am inclined to think that it is only the male." Tame birds of this kind welcome their friends in a similar way.†

"Visitors to zoölogical gardens have probably noticed that the cranes begin to dance when the music strikes up." The one described above danced around his favourite bull. Another made the most ludicrous bounds before a mirror.‡ We can hardly doubt that the various movements described were originally connected with courtship, for they are such as characterize that period in the whole world of birds, but they have apparently become to the crane the expression of general well-being. And since he is so intelligent we may well suppose that he takes pleasure in going through them—that is, that he is playing.

The ostrich struts before his mate with wings unfurled and lowered, sometimes runs very fast, making

---

* See Naumann, ix, p. 362, who also regards these movements as courtship phenomena: so we have here a very clear illustration of their production by association without that sexual excitement which at first must have occasioned them.

† Naumann gives a similar description of the stork, ix, p. 256.

‡ Scheitlin, Thierseelenkunde, ii, p. 76.

three or four sharp turns with inimitable skill, then checks his pace and marches proudly back again, to repeat the sport.

According to Liebe's description of the lapwing, he does not go directly to the female after his exhibition of flying, but makes eyes at her in the funniest way, skipping now to the right, now to the left, and making deep bows with his head held on one side. "At this she will rise and stir about a little and begin a soft twittering which seems to delight her mate, who gives expression to his warmth of feeling by running a few steps nearer and standing while he throws a grass blade or bit of stone behind him, which seems to be the signal for beginning the game anew." Brehm says that the sportive heathcock "holds his tail upright and fan-shaped, his head and neck, on which the feathers are erected, outstretched, and drags his wings. He leaps from side to side, sometimes circles, and finally plunges his bill deep in the ground. The condor spreads his wings, bends his neck stiffly, and turns slowly with little tripping steps and trembling wings. "In North America," says Darwin, "large numbers of a grouse, the *Tetrao phasianellus*, meet every morning during the breeding season on a selected level spot, and here they run round and round in a circle of about twenty feet in diameter, so that the ground is worn quite bare, like a fairy ring. In these partridge dances, as they are called by the hunters, the birds assume the strangest attitudes and run round, some to the left and some to the right." *

I believe I am right in assuming that such dancing motions are not only the means of displaying the colours

* Descent of Man, ii, p. 74.

of the bird's plumage, but, independently of that, produce excitation. If a human example is allowable, the effect of throwing one hip forward is suggestive of what I mean.* That the Greeks understood this is proved by the "line of Praxiteles," which gave to Greek sculpture a certain sensuous charm while preserving its chaste severity.

I pass by the lower animals, though an example given above from fishes seems to indicate that they, too, are playful during courtship. On the whole it seems probable, as I said above, that most of the "courtship arts" were simply excitation reflexes. Since they are influential in stimulating the female, they were favoured by natural selection and rendered constantly more powerful and complicated, until they became full instincts. This is true even in such exceptional cases as those of the butterfly and the spider. It is only to animals with a high degree of intellectual development that we can even hypothetically attribute pleasure in their movements for themselves, the wish to accomplish something, or the desire to make a display, in addition to the habitual reflexes of courtship, so that in the midst of the real exercise of instinct the voice of play would rise as a psychic overtone; or, as James would say, would form a psychic fringe. Such genuine play as that of youth it can not be.

3. *Courtship by means of the Display of Unusual or Beautiful Forms and Colours.*

After what has gone before, it is sufficient to say here that in this case, too, the display is playful only

* Zola furnishes vouchers for this. See especially Nana's appearance in the theatre. There may be some suggestion of this in the pleasure of the waltz.

when the animal making it is intelligent enough to be conscious of self-exhibition. "With mammals," says Darwin, "the male appears to win the female more through the laws of battle than through the display of his charms," * but he adds a long list of sexual stimuli. Nowhere, however, do I recall a description by him or another where a mammal attempted to draw attention to his excited condition by movements, with the single exception of monkeys. Indeed, Darwin says that proof is wanting that the males of mammals make any effort to display their charms to the female.†

But perhaps the actions described in the section on courtship arts, such as the dog's erect carriage, his waving tail and stiff legs, are partly to show his physical advantages, and we read how the stone marten raised his hair, and the fish otter played with his eel-like tail. I have often noticed, too, that dogs who wish to be especially friendly have a way of turning their back to the stranger, which is like the habit of the apes, for the dog often has striking tufts of hair on his hind parts.

We now take up birds again. All the different motions that we have seen described are useful to the bird in displaying his form and colouring. When the reed warbler takes his downward plunge in the air his feathers are inflated till he looks like a ball. The beautiful Madagascar weaver bird flutters like a bat, with trembling wings, about the modest gray female. Naumann says of the blue titmouse: "Hopping busily about in the bushes, swaying on slender sprays, etc., the male dallies with his mate, and at last floats from one tree top to another, sometimes forty feet away, where the

---

* Descent of Man, chap. xvii.

† *Loc. cit.*

widespread wings are not folded and all his feathers are so ruffled up that he looks much larger and is hardly recognisable. But he can not sustain a horizontal flight, and each time sinks perceptibly lower. This kind of floating is not usual with the titmouse, and therefore the more remarkable." *

The hoopoe spreads his fine head decoration in flying as a fan is opened and shut. The striped snipe inflates his feathers and flies slowly with languid strokes, looking much more like an owl than one of his own kind.

The tumbling about in the air common with so many birds, as well as the upward flight and quick descent, also serves to show off their colouring. Dance motions are, however, best of all calculated to display their charms advantageously, and the vanity displayed by many birds during these performances strengthens the probability of self-consciousness. Indeed, when we reflect how early a child shows an appreciation of any expression of admiration, how vain the dog is of his tricks, and the parrot of talking, this supposition does not seem unwarranted. The vanity of peacocks is proverbial. "He evidently wishes for a spectator of some kind," says Darwin, "and, as I have often seen, will show off his finery before poultry, or even pigs."

Gesner remarked, long ago, in his Historia Animalium, that the peacock admired its own beautiful plumage and at once displays his glowing feathers when any one admires them and calls them beautiful. Bennett says the bird of paradise looks knowing and dances about when a visitor approaches his cage. He will not endure the least spot on his feathers, and often spreads his wings and tail to gaze upon his finery. "Espe-

* Naturgeschichte der Vögel Deutschlands, iv, p. 68.

cially in the morning does he try to display all his glory. He busies himself in arranging his plumage. The beautiful side feathers are spread out and drawn softly through his bill, the short feathers disposed to the best possible advantage and shaken lightly, then he raises the splendid long plumes that float like down over his back and spreads them as much as possible. All this accomplished, he runs back and forth with quick bounds, vanity and delight in his own beauty expressed in his every movement. He examines himself from above and below, and gives vent to his satisfaction in loud cries, that are, alas! only harsh noises. After each exhibition it seems to be necessary to rearrange his feathers, but this labour never tires him, and he spreads them again and again, as a vain woman would do."

Let us now notice some birds during courtship itself.

The male *Rupicola crocea*, says Darwin, is one of the most beautiful birds in the world, of a splendid orange colour, and with finely shaped and marked feathers. The female is a greenish brown with red shading, and has a very small comb. Sir R. Schomburgh has described the wooing of these birds. He happened upon a rendezvous where ten males and two females were present. A space of about four or five metres in diameter was cleared as if by human hands, and every blade of grass removed. One male danced to the evident delight of the others; he stretched his wings, raised his head, and spread his tail like a fan, strutting proudly till he was tired and then was relieved by another.

Sometimes a dozen or more birds of paradise collect in full feather, where they hold a "dance meeting," as the natives call it. They flutter about, spread their wings, erect their splendid plumes, vibrating them till, as Wallace remarks, the whole tree top seems made of waving feathers.

Pheasants not only spread and erect their fine crests at such times, but they turn sideways toward the female on whichever side she happens to be standing, and incline the beautiful outspread tail in the same direction. When a peacock wishes to make a display he stands opposite the female, spreads his tail and raises it perpendicularly, at the same time showing to advantage his beautiful neck and breast. Those, however, that have dark breasts and eye marks all over the body display their tails a little diagonally and stand in such a position that the eye marks are clearly seen by the female. In whatever direction she turns the outspread wings and tail held diagonally always confront her.

It seems undeniable that there is in this kind of courtship a conscious display of personal charms, and therefore play.*

Following Darwin's account, we now turn to birds of more sober plumage. The bullfinch approaches his mate from the front, inflating the brilliant red feathers on his breast so that they are much more conspicuous than usual, and twisting his black tail in a comical manner. The common linnet inflates its rose-coloured breast and spreads its brown wings and tail, showing the white border to the best advantage.† The goldfinch behaves differently from other finches. His wings

* Darwin, Descent of Man, ii, pp. 82–94; see also, on p. 98, the strange behaviour of the Argus pheasant (described by T. W. Wood).

† Darwin, however, cautions us against supposing that spreading the wings is solely for the purpose of displaying the colouring.

are fine; black shoulders with dark pointed quills picked out with white and gold. Weir confirms Darwin's statement that no other British finch turns from side to side as he does in courtship, not even the close-related siskin, for it would not enhance his beauty. The common pigeon has iridescent breast feathers, and there fore inflates them, but the Australian pigeon (*Ocyphaps lophotes*), that has handsome bronze wings, acts quite differently. Standing before the female he sinks his head almost to the ground, raises his widespread tail, and half opens his wings, then he lets his body rise and fall with a slow motion that causes the glittering feathers to shine brilliantly in the sunshine.*

Karl Muller tells us that the red wagtail prostrates himself at the feet of his bride, flapping his wings and dragging the outspread fan of his tail on the ground. The crossbill perches on the highest limb of the tallest tree, singing lustily and whirling about incessantly the while. The snipe ardently draws near his mate with inflated feathers, lowered wings, and tail raised and spread When the cuckoo feels the stirrings of love he "inflates his throat feathers, hangs his wings, moves his partly spread tail up and down, turns from side to side and bows to his lady as often as he cries 'Cuckoo.'" † The orange bird pursues his mate in apparent wrath and then bows and scrapes before her. Brehm describes the pairing of golden-crested wrens very beautifully: "The male inflates his crest until it forms a splendid crown, in which the black stripes extend far down the side of his head without concealing the white eye marks and displaying the flame-coloured parting most advantageously." ‡

* Descent of Man, *loc. cit.* † Naumann, v, p. 216.
‡ Chr. L. Brehm, Beiträge zur Vögelkunde, ii, p. 138.

Even if one fully agrees with Mr. Wallace that sexual selection does not actually create the beauty of bird plumage, it is hardly possible in the face of facts like these to deny that at least those developments of colour and other ornamentation which transcend the uses of concealment and warning must have some connection with the sexual life. The substitution of unconscious for conscious choice makes this connection clearer, but the acceptance of the theory herein previously developed—namely, that of the importance to race life of feminine coyness and the necessity on the part of males to overcome it by such means—does away with all choice, and relegates the whole subject to the sphere of natural selection.

4. *Courtship by means of Noises and Tones.*

Here, too, the view set forth in the last section is applicable. The ordinary sounds emitted by the excited male probably have the same effect as the husky voice and laboured breathing of civilized man * They furnish material for the working of selection in the production of courtship arts which are later used also for other purposes. Among the higher animals imitation often plays a part as important as that of selection—indeed, it sometimes supplants the latter in cases where hereditary courtship arts are rudimentary only, each individual acquiring the finer points by imitation. We may suppose, for example, that many young birds learn from their elders that they must fight for a mate, and in turn teach it to the next generation. By this method an art would be acquired founded on an instinctive basis, but not in-

* The writers of a certain class of modern novels excel in the portrayal of such phenomena.

herited by the individual—similar, *cum grano salis,* to the fine arts of savages. For by outgrowing instinct they reach, through teaching and imitation, a certain degree of development to which they remain constant so long as the conditions remain constant, but would at once fall back to the level of hereditary instinct were the individuals to lose their model. The more important the part play by imitation, the more probability of a playful expression of the activity in question.*

I again pass over the lower orders, although they offer much that is of the greatest interest. There is very little that deserves to be called vocal art in the courtship of mammals; most of them confine their acoustic demonstrations to a passionate howl, roar, shriek, or growl, or to the simple call. The performance of howling apes, however, is a notable exception, for they collect in companies and frequently give concerts that last for hours Hensel says: "In summer, when the beams of the morning sun have dispelled the night mists, the howling apes leave the shelter of the thickly leaved trees to which they have clung all night. After satisfying their hunger they have time before the heat of the day to indulge in social pleasures which, as befits animals so serious, are free from the unseemliness that characterizes those of many of their relations. They now repair to the shelter of some gigantic monarch of the forest whose limbs offer facilities for walking exercises. The head of the family appropriates one

* See Weismann in the Deutschen Rundschau, October, 1889, p. 63: "A young finch, brought up alone, sings untaught the note of his kind, but never so well as those that have had the advantage of parental example. He, too, is governed by a tradition; the essentials only of the finch's song are inherent in his organism, are born in him."

of these branches and advances along it seriously, with elevated tail, while the others group themselves about him. Soon he gives forth soft single notes, as the lion likes to do when he tests the capacity of his lungs. This sound, which seems to be made by drawing the breath in and out, becomes deeper and in more rapid succession as the excitement of the singer increases. At last, when the highest pitch is reached, the intervals cease and the sound becomes a continuous roar, and at this point all the others, male and female, join in, and for fully ten seconds at a time the awful chorus sounds through the quiet forest. At its close the leader begins again with the detached sounds." How can we explain this strange concert? This description gives the impression that it is merely a social game, but how did the animal acquire the instrument on which he plays, the throat thickened as with a gôitre? A. von Humboldt says: "The small American monkey chirps like a sparrow, having simply an ordinary hyoid bone, but that of the great ape is a large bony drum. The upper part of the larynx has six compartments, in which the voice is formed. Two of these compartments are nest-shaped and very like the lower larynx of birds. The doleful howl of the ape is caused by the air streaming through this great drum, and when we see how large an instrument it is we are no longer surprised at the strength and range of this animal's voice, or that it gives him the name he bears." Such a structure as this must serve some useful purpose, and the idea of courtship suggests itself as the probable use, in the first instance, since it outweighs all other causes for excitement Then its exercise may have come by association to be purely playful.

Scheitlin says of the cat: "Their pairing time is

interesting. The male is coy, the females who visit him sit around him while he growls in a deep base. The others sing tenor, alto, soprano, and every possible part as the chorus mounts, constantly growing wilder. They shake their fists in one another's faces, and will not let even him whom they have come to visit approach them. On clear moonlight nights they make more noise than the wildest urchins." This certainly seems something more than mere sportiveness, and must unquestionably be set down as connected with courtship. Darwin regards the cry of the howling ape in the same light, and in addition has this to say about the *Hylobates agilis:* "This gibbon has an extremely loud but musical voice. Mr. Waterhouse states: 'It appeared to me that in ascending and descending the scale, the intervals were always exactly half tones, and I am sure that the highest note was the exact octave to the lowest. The quality of the notes is very musical, and I do not doubt that a good violinist would be able to give a correct idea of the gibbon's composition, excepting as regards its loudness.' * This gibbon is not the only species in the genus which sings, for my son, Francis Darwin, attentively listened in the zoölogical gardens to a *Hylobates leuciscus* which sang a cadence of three notes in true musical intervals and with clear musical tones. It is a more surprising fact that certain rodents utter musical sounds. Singing mice have often been mentioned and exhibited, but imposture has commonly been suspected. We have, however, at last a clear account by a well-known observer, the Rev. S. Lockwood, of the musical powers of an American species, the *Hesperomys cognatus,* belonging to a genus distinct from

* Darwin adds that Owen confirmed this observation.

that of the English mouse. This little animal was kept in confinement, and the performance was repeatedly heard. In one of the two chief songs 'the last bar would frequently be prolonged to two or three, and she would sometimes change from C sharp and D to C natural and D, then warble on these two notes a while and wind up with a quick chirp on C sharp and D. The distinction between the semitones was very marked and easily appreciable to a good ear.'" *

Coming again to birds we first note their characteristic song. Brehm and Lenz tell us of finches: "Their song is called a strophe because it consists of one or two rhythmic measures, given with great persistence and sometimes with rapidity. To this the finch owes its popularity among fanciers, who distinguish a great number of such strophes and give them each a name until their study has become quite a science, involved in much mystery to the uninitiated; for while there is little difference between them to the unpractised ear, these people distinguish twentyor more distinct strophes. According to Lenz, one kind of snipe has nineteen strophes when he is free. The syllables of a good double strophe are as follows: Zizozozizizizizizizizizirreuzipiah totototototozissskutziah The nightingale's song contains from twenty to twenty-four distinct strophes, and according to Naumann's fine description, "is characterized by a fulness of tone, a harmony and variety that are found in the song of no other bird, so that she is rightly called the queen of songsters. With indescribable delicacy, soft flutelike notes alternate with trembling ones, melting tones with those that are joyful, and melancholy strains with ecstatic outbursts. If a soft note begins

* *Cf.* Darwin, Descent of Man, ii. p. 263.

the song, gaining in strength to the climax and then dying away at the end, the next strophe will be a series of notes given with hearty relish, and the third a melancholy strain melting, with purest flute notes, into a gayer one. Pauses between the strophes heighten the effect of these enchanting melodies; they and the measured *tempo* must be noted, fully to comprehend their beauty. We are amazed at first at the number and variety of these bewitching tones, then at their fulness and power coming from a creature so small. It seems almost a miracle that there can be such strength in the muscles of its tiny throat" * Bechstein has attempted to write the syllables of its strophes thus: †

Tiuu-tiuu-tiuu-tiuu,
Spe, tiuu, squa,
Tio, tio, tio, tio, tio, tio, tio, tix,
Qutio qutio qutio qutio,
Zquō zquō zquō zquō,
Tsu, tsü, tsu, tsü, tsü, tsü, tsü, tsü, tsü, tsi.
Quoror, tiu, zqua, pipiqui,
Zozozozozozozozozozozozo Zirrhading!
Tsisisi tsisisisisisisisi,
Zorre, zorre, zorre, zorre hi;
Tzatn, tzatn, tzatn, tzatn, tzatn, tzatn, tzatn, zi,
Dlo, dlo, dlo, dlo, dlo, dlo, dlo, dlo, dlo,
Quio tr rrrrrrrr itz
Lu lu lu. ly, ly, ly, lî lî lî,
Quio, didl li lulyli
Ha gürr, gürr, quipio!
Qui, qui, qui, qui, qi qi qi qi, gi gi gi gi;

* Naumann, ii, p. 381.

† J. M. Bechstein, Naturgeschichte der Stubenvögel, p. 321.

Gollgollgollgoll gia hahadoi,
Quigi horr ha diadiadillsi!
Hezezezezezezezezezezezezezezezezeze quarrhozehoi;
Quia, quia, quia, quia, quia, quia, quia, quia ti:
Qi qi qi jo jo jo jojojojo qi—
Lu ly li le là la lo lo didl jo quia
Higaigaigaigaigaigaigai gai gaigai,
Quior ziozio pi.*

Thrushes, unlike most birds, sit still when they sing, and the songs, too, have a soothing quality. They choose the summit of tall trees for their perch, as if to avoid interruption.†

The song of the blackbird that perches, on fine evenings, on the topmost gable of a roof or the very highest branch of a tree and lifts his deep and yet clear and joyous voice is perhaps the most æsthetically effective of all. Audubon says of the cardinal bird: "His song is at first loud and clear and suggestive of the best tones of a flageolet, but it sinks lower and lower until it dies away entirely. During his love time this noble singer produces his notes with more force, and seems conscious of his strength; he swells his breast, spreads

---

* Naumann found quite a different song common in his neighbourhood, and, indeed, the nightingale's song varies very much, which goes to prove that in so highly developed an art tradition and imitation play an important part. But individual differences, too, are found in their songs and those of the thrush and other birds. For older imitations of the nightingale, see O. Keller's Thiere des classischen Alterthums, p. 317.

† [So also does the American mocking-bird, often choosing the tip of a lightning-rod. As the song proceeds the notes come faster and faster, until the bird is lifted off the perch, thrown fluttering straight up in the air, sometimes to a height of three or four feet, and falls again by somersaults to the perch, never stopping the song.—J. Mark Baldwin.]

his scarlet tail, flaps his wings, and turns from side to side as if he would express his joy in the possession of such a voice. Again and again the song is repeated, the bird only pausing to get breath." Brehm relates of the whistling and the scarlet shrike: "The most remarkable thing about these birds is undoubtedly the use they make of their song, which is, properly speaking, not a song at all, being but a single strain, sonorous as few often repeated notes are, and common to the two sexes. The call of the former consists of three, rarely two, distinct sounds, pure as a bell and all within the octave, beginning with a moderately high note, followed by a deeper one, and concluding with one still higher. These, like the piping of the scarlet shrike, are peculiar to the male bird, but his mate answers at once with an unmusical cackle or chick which is difficult to imitate or describe. The female scarlet shrike only begins her cackle when her mate has finished his call, but the whistling shrike usually joins him on the second note, but both show a surprisingly quick ear, and never keep him waiting. Sometimes she cackles three or four or even six times before the male joins in, but when he does so the whole performance begins over and proceeds in regular form. Several experiments have proved to me that the two sexes always act together. I have killed now a male and now a female to make sure. When either falls and is, of course, silenced, the other anxiously repeats the call several times." The Prince von Wied says: "The bell bird, both by reason of its splendid white plumage and its clear, loud voice, is one of the attractions of a Brazilian forest, and is usually noticed at once by a stranger. His cry resembles the tone of a very clear bell, sounded once and then withheld for a long interval, or at times repeated rapidly, in which case it is like a blacksmith's strokes on the anvil." *

Brehm carefully observed one of these bell birds in captivity, and describes minutely its wildly excited condition, which becomes more and more intense as the cries are repeated: "That he sometimes even seals these love transports with his death is proved to me by the fact that the bell bird which I was watching fell dead from his perch with his last cry." One can hardly say in this case that the birds sing from mere exuberant spirits. Other birds show similar ecstasies, notably the black and heath cocks. The voice of the former is exceedingly high and is indescribable in words; their cry is well known to hunters and is commonly heard in the spring. About sundown this bird perches on a tree, preferably an old beech or fir, that he will return to year after year if not disturbed. At the time when the red beech leaves he sings with only short intermissions from the first gleam of dawn till after sundown. He takes his post on a bare, sturdy limb, inflates his long neck feathers, makes a wheel of his tail, drops his wings, erects his plumage, trips on his toes, and rolls his eyes comically. At the same time he gives forth notes that are at first slow and detached, then quicker and more connected, until at last a distinct beat can be distinguished among the accompanying notes, ending in a long-drawn cry, during which the bird rolls his eyes in ecstasy." †

* This bird has a bill of the most peculiar construction. It has a flaccid bag hanging below it that is inflated during courtship sometimes to a length of three inches, and a perpendicular position. It would be difficult to explain the purpose of this appendage if one did not admit its connection with the sexual life. Romanes gives a description in his Darwin and after Darwin.

† F. von Tschudi, Das Thierleben der Alpenwelt, p. 174.

I need not multiply these examples. Enough has been said to show that birds invariably sing during their mating time, but not exclusively then. The blackcock, starling, and robin also sing out of this season, as well as the water ouzel, which Tschudi has so beautifully described, while the wren, red linnet, and goldfinch can be heard all winter; the white-throat, too, sings all the year round. Indeed, it may be that the breeding time of some birds is variable, as seems to be the case with the water ouzel, which, Tschudi says, "does not confine itself to any particular month; the young just hatched may be seen even in January." Besides, birds sing not only before pairing but all through the breeding time; in numberless cases the male pours out his sweetest song while his mate is on the nest. This is obviously play, rather than courtship. The duets * that they sometimes produce together are probably the effect of heredity; while in other cases the male song is taken up by imitation on the part of the female. Hudson says that a singing female usually has plumage the same as her mate.† Finally, there are rare cases where the male sings better at other times than during his courtship. Spencer, in his article "On the Origin of Music," says this is true of the thrush.‡ And Hudson says of a small yellow finch found in La Plata that in August, when the trees are blooming, a flock of these birds will appear in a plantation, perching on the boughs and beginning a concert in chorus, "producing a great

* Examples in Darwin's Descent of Man, chap. xii.

† The Naturalist in La Plata, p. 283.

‡ Mind, xv (1890), p. 452 Spencer sees in this fact a contradiction of Darwin's theory. I do not understand why, since it is so probable that vocal reflexes in general are transmitted by heredity, and so may always be called forth by other excitation.

volume of sound as of a high wind when heard at a distance," and this takes place daily for hours at a time. But during his courtship the male has but "a feeble, sketchy music," regaining his skill only after the nest is built. This is a more valuable example than Spencer's, for he observed only a single bird that may have been sick at the time for pairing, while Hudson's observation refers to a whole species. Yet the phenomenon is too rare to have any weight against the overwhelming mass of evidence for the view that song in general belongs to courtship. It is wiser to seek some special explanation of these irregular cases, and also to bear in mind that "better" and "worse" are relative terms. A song broken by the restless motions of an excited bird may seem not so good to the listener as the same strain produced when the singer is quiet and his notes are therefore louder and more continuous There is also a possibility that song is sometimes supplanted by the disproportionate evolution of other courtship arts—the finch spoken of by Hudson has unusual powers of flight and skill in dancing. However, I do not profess to find an adequate answer in these suppositions to this undeniable difficulty.

Those instances in which the bird expresses his excitement by means of a kind of instrumental music, instead of doing it vocally, are also very remarkable. Darwin has a long series of such examples. Peacocks rattle the quills of their tails, and birds of paradise do the same thing, during their courtship. Woodpeckers call the females by striking the bill very rapidly on dry wood, making in this way a sort of drumming sound Turkeycocks scrape their wings on the ground. Many birds make a kind of whirring sound in flight; a familiar instance is the "beating" of army snipes as they mount rapidly aloft in the evening. It is evidently a call to the female, who answers from the earth with a "dick-küh" or "kup ti küpp ti küpp." *

Naumann thought the flapping of storks connected with courtship, but as I do not consider these manifestations playful I abstain from further citations, except in the case of the bittern, which may be said to practise his art playfully if the following description is to be trusted. Brehm says: "The peculiar pairing call of the male bittern is like the lowing of oxen, and on still nights may be heard at a distance of two or three kilometres It is composed of a prelude and a principal tone, and sounds something like 'Ueprúmb'† at a distance. It is said that on coming near the birds a sound like beating on water with sticks is heard. . . . The male keeps it up almost constantly; beginning at twilight he is most vociferous before midnight, and ceases at dawn, only to start up again, however, between seven and nine o'clock. The observations of Count Wodzicki have confirmed the account of the older writers. He says: 'The performer stands on both feet with his bill in the water when giving vent to this extraordinary sound, which causes the water to spurt up all around. First I heard Naumann's 'Ue' and then the bird raised his head and looked behind him, but quickly plunging it in again he produced such a roar that I was startled. I am convinced that these tones which

* Diezel's Niederjagd, p. 664. Here, too, is to be found a history of the controversy over the origin of this sound, but unfortunately Darwin's remarks on similar phenomena are not noticed.

† Although he often attempted it, Naumann never got a sight of the bittern. We may assume that his "Ueprúmb" is only a rough approximation of the sound, which is not transcribable. "It is a sound," say the Müllers "that blends the deepest lowing of cattle with water splashing and something like sighs."

are loudest at their beginning are produced when the bird has his throat full of water and expels it with great force. The music went on, but he did not throw his head back again, nor did I hear the loud note any more. It seems to express the highest pitch of excitement, and having given vent to it he is relieved. After an interval he cautiously raised his bill from the water and peered around, for it seems that he can not tear himself away from his charmer.' * The bittern stands in an open space, where the female can see him during his performance. The splash is caused by his striking the water several times with his bill before plunging it in; other water sounds are produced by the falling drops, and the last one by the emission of what remains in his bill. A male disturbed by Wodzicki flew off and spurted out a considerable stream that had collected in this way."

5. *Coquetry in the Female.*

I have attempted, in the theoretical part of this chapter, to show that the instinctive coyness of females is the most efficient means of preventing the too early and too frequent yielding to sexual impulse A high degree of excitement is necessary for both, but the female has an instinctive impulse to prevent the male's approach, which can only be overcome by persistent pursuit and the exercise of all his arts. This coyness often seems like fear, and sometimes even like anger, as in the case of spiders and preying animals, but sometimes there is no fear at all, the animal even inviting the male's approach until he shows some eagerness, then

* The happy female keeps near her mate, in a crouching position and with erected crest and half-shut eyes, as if bewitched by his boisterous wooing (Müller, Thiere der Heimath, ii, p. 469).

her coquetry manifests itself in alternate calling and fleeing. It is not essentially playful, for it is a struggle between opposing instincts and has a serious object, but we can easily see how it becomes play when unconnected with the strong emotions of fear or anger—that is, when it is a sort of kittenishness. Then the flight and resistance of the female, though they are not play pure and simple, take on something of the character of a game and temper the rough force of instinct.

As adequate descriptions of such playful coquetry are rare, I have only a few examples from the higher animals. The Mullers describe as follows the gambols of a pair of squirrels: "The male comes near and flees, grunts and whispers, runs and leaps, approaches his mate and leans against her; she turns away and lures him on, appears indifferent and then tries to please him, changes from momentary anger to frisky good humour; the bounds and chase go on so rapidly that one can scarcely follow their turns, and finds himself charmed by the sight of this artless sportiveness, as graceful as it is beautiful." * "Another exquisite game may be seen in April and May, when the pairing watershrews carry on their teasing chase The fleeing female pretends to hide, crouching in mole holes and under stones, roots, and rubbish while her mate looks for her. Or she skips out, throws herself in the water, runs across on the bottom and clambers to a new place on the other side of the brook; but he soon spies her and follows in her footsteps. So the game goes on, with only rest time enough for them to eat in." †

The doe, in her breeding time, calls to the buck in clear tones that bring him to her side at once, then she,

* Thiere der Heimath, vol. i, p. 196. † Ibid., p. 280.

"half in coyness, half in mischief, takes to flight at his eager approach, makes toward an open space, and runs in a circle. The buck naturally follows, and the chase grows hot and as exciting as a race of horses on a track. To the frequent high calls of the fleeing doe are added the deep, short cries of the panting buck; but suddenly the roguish doe disappears like a nymph into the thicket near at hand, and the baffled buck stands with head erect and ears thrown forward; then we see his head lowered as he catches the scent, and he too vanishes in the wood." *

It is a familiar fact that female birds must be long courted and pursued before they yield. L Buchner has collected some examples proving this.† Mantegazza says: "Coquetry is not the exclusive prerogative of the human female. No woman ever born could surpass the abominable (!) refinement of cruelty displayed by a female canary in her pretended resistance to her mate's advances. All the countless devices of the feminine world to hide a Yes under a No are as nothing compared with the consummate coquetry, the deceptive flights, the bitings, and thousand wiles of female animals."

However mistaken the conclusions here drawn from this antagonism of sexual impulse and coyness, the fact undoubtedly remains that coquetry is exceedingly widespread among birds. Thus the female cuckoo answers the call of her mate with an alluring laugh that excites him to the utmost, but it is long before she gives herself up to him. A mad chase through tree tops ensues, during which she constantly incites him with that mocking call, till the poor fellow is fairly driven crazy.

* Thiere der Heimath. vol. i. p. 429.

† Liebe und Liebesleben in der Thierwelt, pp. 89 f.

The female kingfisher often torments her devoted lover for half a day, coming and calling him, and then taking to flight. But she never lets him out of her sight the while, looking back as she flies and measuring her speed, and wheeling back when he suddenly gives up the pursuit. The bower bird leads her mate a chase up and down their skilfully built pleasure house, and many other birds behave in a similar way. The male must exercise all the arts that have been described in these pages and more before her reluctance is overcome. She leads him on from limb to limb, from tree to tree, constantly eluding his eager pursuit until it seems that the tantalizing change from allurement to resistance must include an element of a mischievous playfulness.

CHAPTER V.

THE PSYCHOLOGY OF ANIMAL PLAY.

ALTHOUGH the mental accompaniments of play have often been referred to in the preceding chapters, that mention was but cursory, and it is necessary, in summing up, to consider them more fully. First, then, let us recall the position reached in the first two chapters, where the play of young animals formed our principal problem. We there said that if this should be explained satisfactorily, then adult play would not offer any great difficulty, an assumption warranted by the fact (treated of in the third chapter) that all genuine play is at first youthful play. Even love play, which as we have found can hardly be said to be genuine play, appears in early youth, and when the word play is applied to the acts of grown animals at all it is chiefly with reference to those that are experimental—namely, to games of motion, which are really child's play furnishing practice for the later exercise of important instincts.

For adult animals which are already practised in their plays, the Schiller-Spencer theory of surplus energy may apply, though experience of the pleasurableness of play gained in youth is of great importance too. But in youthful play the biological significance of the phenomenon—namely, that it relieves the brain from the finely elaborated hereditary tracts and so furthers intellectual development—becomes much more prominent than the merely physiological. Indeed, we found it probable that surplus physical energy is not even a *conditio sine qua non*, for in youth the instinct for playful activity is urgent even when there is no surplus of energy. In following out this idea the psychological aspect of the question was touched upon only incidentally, and we found the essential point in the definition of play to be its quality of practice or preparation, either with or without higher intellectual accompaniments, in distinction from the serious exercise of instinct. This is a great advance in so far, but then we often do not know whether even a child is conscious that it is only playing. So it is time to inquire in what the mental accompaniments of play consist, when they are present, and it is apparent from the nature of the question that its answer must be sought in the emotional life.

The feeling of pleasure that results from the satisfaction of instinct is the primary psychic accompaniment of play. There are, indeed, instincts whose exercise is connected with decidedly disagreeable feelings; but instinctive activity as such is usually pleasurable, when psychic accompaniments are present at all. If we accept A. Lehmann's definition of pleasure as a state of temporary harmony between psychic and physical life conditions,* we may be sure of its presence in most instinctive activity not marred by the emotions of anger or fear which are sometimes prominent. Since these hindrances are not operative in play, and since also the power of instinct is here exceptionally strong,† we may

* A. Lehmann, Die Hauptgesetze des menschlichen Gefühlsleben, 1892, p. 150.

† P. Souriau (Le plaisir du mouvement, Revue Scientifique, xvii, p. 365) says the need of movement is especially great in the

safely assume that strong feelings of pleasure accompany it.

And, further, energetic action is in itself a source of pleasure. Experiments made with the dynamometer, sphygmograph, pneumatograph, and plethysmograph show that pleasure is accompanied by strengthened muscular activity, quickened pulse-beat and respiration, and increased peripheral circulation. It is not strange, then, that the energetic activity of play with its analogous physical effects is connected with feelings of pleasure. P. Souriau says: "When we indulge in exercise that requires the expenditure of much energy all our functions are quickened, the heart beats more rapidly, respiration is increased in frequency and in depth, and we experience a feeling of general well-being. We are more alive and glad that we are." * Very rapid and lively emotions produce "a sort of intoxication and giddiness that are most delightful." Besides these external effects of pleasurable feelings they are accompanied internally by a heightened excitation of the sensor and motor centres of the cerebrum, much like that produced by concentrated attention—a fact which points to the probable explanation of the physiological side of pleasure by means of the only purely intellectual play of animals, curiosity.

The unconscious connection of emotional accompaniments with intellectual activity is shown still more clearly in that joy in ability or power which has confronted us as the most important psychic feature of play throughout this whole treatise.

---

young animal, "because he has to try all the movements which it is necessary for him to make in later life."

* Ibid.

This feeling is first a conscious presentation to ourselves of our personality as it is emphasized by play—a psychological fact which Souriau states in the words "We are more alive, and glad that we are." But it is more than this, it is also delight in the control we have over our bodies and over external objects. Experimentation in its simple as well as its more complicated forms is, apart from its effect on physical development, educative in that it helps in the formation of causal associations. Knowledge of these is arrived at first by means of voluntary movements, and afterward extended in various directions,* and playful experimentation is a valuable incentive to such movements. The young bear that plays in the water, the dog that tears a paper into scraps, the ape that delights in producing new and uncouth sounds, the sparrow that exercises its voice, the parrot that smashes his feeding trough, all experience the pleasure in energetic activity, which is, at the same time, joy in being able to accomplish something.

But what is this feeling of joy, in its last analysis? It is joy in success, in victory. Nietzsche has opposed the "struggle for power" to Darwin's "struggle for existence," and however contradictory it may seem to identify the survival of the fittest, which is usually no struggle at all, with a struggle for power, it is certain that striving for supremacy is instinctive with all intelligent animals.

The first object to be mastered is the creature's own body, and this is accomplished by means of experimental and movement plays. This achieved, the animal's spirit of conquest is directed toward inanimate objects, and very easily degenerates into destructiveness. But he

* See Sully, The Human Mind, vol. i, pp. 264, 444; vol. ii, p 224.

aspires still higher, and attacks other animals in playful chase and mock combats; the fleeing animal will playfully escape from his pursuer. In the other forms of play—building, nursing, and curiosity—the impulses of ownership and subjugation manifest themselves in various ways. Imitative play is full of rivalry, and it is a powerful motive in courtship. It is a satisfaction that can not be attained without effort, and is increased in proportion to the difficulty of overcoming opposition, without which there would be no consciousness of strength. This is just as true in simple muscular co-ordination as in the solution of the problems of a game of chess.

In short, we see in this joy in conquest a "correlative to success in the struggle for existence," * whether it concerns rivalry among comrades, victory over an enemy, the proof of one's capabilities, or the subduing of an external object.

In view of all this, it seems a very mistaken proceeding to characterize play as aimless activity, carried on simply for its own sake. Energetic exertion may be provocative of pleasure, as we have seen, but it is by no means the only source of the pleasure produced by play. "Disinterested play!" exclaims Souriau in the passage already cited from—"to talk about such a thing is to expose our ignorance. Players are always interested in the result of their efforts." It may be an insignificant aim that inspires us, but there is always some goal that we are striving for, an "end to attain," whose value our imagination usually enhances. "Tell me, if you will, that I am voluntarily deceiving myself; tell me even that I am making myself the dupe

* Spencer, Principles of Psychology, vol. i, p. 534.

of a conscious illusion. It is true, all the same, that activity for its own sake is not enough for me, and I am not interested in a game unless it excites my *amour propre*. I must have a difficulty to overcome, a rival to surpass, or at least be able to make progress." * Grosse says the same thing: "Play stands as a connecting link between practical and æsthetic attainment. It is distinguished from art by the fact that it strives constantly for the attainment of some external aim, and from work in that its satisfaction arises not from the value of its results, but from the achievement itself" The relation of the three can be illustrated by calling work a line, play a spiral, and art a circle." †

While these passages are conclusive as to the fact that play should never be characterized as aimless activity, Grosse's utterance might very easily give rise to false generalizations. In my opinion, adequate psychological definitions of work, play, and art are not to be produced with such "neatness and despatch" as Grosse attempts.

Play is easy enough to define objectively, as practice in distinction from the exercise of important instincts. But in regard to its psychological accompaniments in the playing subject the case is different Here we must suppose a progressive development from mere satisfaction of instinctive impulse (where the act is performed neither for its own sake nor for the sake of an external aim, but simply in obedience to hereditary propensity) through what is subjectively considered akin to work, up to make-believe activity with an external aim as its second stage. Finally, as the outward aim gives way

* *Cf.* K. Lange's Bewusste Selbsttauschung.

† E. Grosse, Die Anfänge der Kunst, 1894, p. 47.

before the pleasure-giving quality of the act itself, the transition to art takes place. At this point the outward aim has but a very slight significance, though never vanishing entirely; for it can not be denied that in artistic execution it regains very considerable importance in an altered form.

Let us take an example that follows all these developmental stages. If a very young puppy is tapped on the nose with the finger, he snaps at it. This is a playful expression of the fighting instinct, where the propensity to obey hereditary impulse is the sole cause for the act, since neither feeling nor an idealized external aim can be alleged as such; it is clearly a reaction to stimulation without higher psychic accompaniments. Going a step further, we will suppose a young dog that chases his brother for the first time and seizes him by the throat. Here the most probable supposition is that subjectively there is no difference between practical activity and this kind of play. The dog has the serious purpose to take the skin in his teeth, to throw his comrade and hold him fast on the ground. It is altogether improbable that he is making believe at first. Here, then, play appears psychologically as quite serious activity, and a little attention to the subject will show that this is a very common condition among human beings.* In the third stage the dogs are grown larger and can bite effectively if they choose; nevertheless, they seldom hurt one another in their tussles. A consciousness of make-believe is rising gradually, and to the force of instinct is being added the recollection of the pleasure-giving qualities of play.

* For instance, a little girl two or three years old will seriously try to feed her doll with soup, or beat it severely. And how many billiard or chess players take defeat seriously!

Only in this way can we explain the animal's restraining his fighting propensities, beyond a certain limit, though the external aim, the subjugation of his opponent, remains and tries hard to break through these restraints. Now, the full-grown dog romps with his master and the make-believe is fully developed and conscious, for his bite is intentionally only a mumbling, his growl pure hypocrisy. The animal, playing a part as an actor, comes very near to art, henceforth he plays for play's sake with very little external aim, though his disposition to use his strength in earnest as the play grows more exciting, witnesses to the fact that it has not entirely vanished. At this point of the illustration we go to man for our instance.

Suppose instead of the dogs two boys wrestling; here, too, we find the earnest aim to overcome an opponent, and at the same time consciousness that the pleasurable quality of the game can only be preserved by confining the struggle to certain limits and keeping up the pretence. Going on to a wrestling match before spectators the case is much the same, for the likeness to real fighting gained in one way is lost in another, since the most reckless wrestlers are held in check by external restrictions, called "rules of the game." Going on further, we make a great advance if we allow the contestants to arrange it all in advance: "You take a good grip and throw me, but I make a sudden move and get the upper hand," and so on. This, then, becomes pure make-believe, since both wrestlers are playing a part; but we shall find that, just as with the dogs romping with their master, the real aim of conquering an opponent will get the better of these restrictions if a particularly skilful move calls forth loud applause. But to go on with the illustration. Supposing the game carried out according to agreement, is all outward aim done away with? By no means. It reappears in a modified form, in the desire to impress the hearers or spectators, and is at bottom our familiar pleasure in power, delight in being able to extend the sphere of our ability, a motive which should never be underestimated, Even the artist does not create for the mere pleasure of it; he too feels the force of this motive, though a higher external aim to him is the hope of influencing other minds by means of his creations, which, through the power of suggestion, give him a spiritual supremacy over his fellow-creatures. This suggestive effect is his real aim, for while it is true in a sense that the artist should not regard applause by the multitude, but listen rather to the voice in his own breast, it is yet nonsense to say that a great artist has no thought of the effect on others.* What is nobler or more kingly than to rule by natural right? Spiritual supremacy is the aim of the highest art, and there is no real genius without the desire for it.

So we find in this pleasure in the possession of power the psychological foundation for all play which

* Grosse is much too clear a thinker not to recognise this In his "scaffolding" of definitions he has this sentence: "Æsthetic effort is not a means to an end outside of itself, but is its own end" (p. 46). But soon after he says: "The artist works not for himself alone, but for others; and if it is too much to say that he creates solely with a view to influencing others, it is yet true that the form and trend of his effort are determined essentially by his conception of the public whom he addresses. A work of art always reveals as much of the public as of the artist, and Mill was guilty of a serious mistake when he said that the characteristic quality of poetry is that 'the poet never thinks of a hearer' On the contrary, the poet would probably never give expression to his thoughts if there were no hearers" (p. 47).

has higher intellectual accompaniments. But it should be remarked that the pleasure is greater when the action involves movements that are agreeable to the senses. Souriau finds an important source of pleasure in movements that overcome resistance. In many movement plays the earth's attraction is the opponent we seek to conquer. The rapid horizontal movement, the leap, the forward motion of a swing, are a mock victory over the force of gravitation. This is a most pregnant idea, and doubtless true essentially, though there is a difficulty in the fact that the backward motion of the swing, the leap into water, and the lightning speed of coasting and skating, all of which depend on the untrammelled action of gravitation, are just as pleasurable. The downward flight of birds, so often referred to in this book, belongs to the same category. Still, this does not disprove Souriau's idea, for, while weight is not actually overcome in these exercises, there is freedom from all the unpleasant effects of weight, such as friction, jarring, etc. All gliding, slipping, rocking, and floating motions give us a peculiar and agreeable feeling of freedom, whether they are contrary to gravitation or not. We are freed from all the little jars and rubs that usually accompany our motions, and are primarily the effect of weight; hence these gliding motions are particularly agreeable to the senses and tend greatly to increase the pleasurableness of play. The same is true of agreeable sounds and colours when they have place in a game.

If pleasure in the possession of power appears as the most important psychological foundation of play, its highest intellectual expression, its idealization, as it were, proves to be the assuming of a rôle or mock activity in any form. Objectively all play is of this character, since it employs an instinct when its actual aim is wanting, but subjectively play is not always sham occupation. It is safer, as we have seen, to assume that the primary forms have none of this. Only when the chase and fighting plays have been so frequently repeated that the animal recognises their pleasurable quality, can we assume that even an intelligent creature begins consciously to play a part. We may be quite sure of it, however, when he uses his weapons guardedly and shows signs of friendship to his opponent, or when he tosses a bit of wood in the air and catches it again. As regards other kinds of play we are only justified in thinking it *probable* that such a consciousness of shamming is present; that monkeys, for instance, labour under a kind of mock excitement when they indulge their destructive impulses, and that the bird tumbling about in the air has some object when he seems on the point of falling helpless to the earth; that the parrot that knocks on his cage and cries, "Come in!" is consciously making believe; that the wooing bird really *plays* the agreeable, and that his mate coquettes intentionally, etc.

But in case the making believe can not always be established, it is useful to remember that actual deception is not rare among the higher animals. Any one who has had much to do with dogs will not doubt for a moment that this is true. I once saw one drop a piece of bread that he would not eat, on the ground and lie down on it, then with an air of great innocence pretend to be looking for it The Müllers tell of a pointer that shammed sleep after he had licked all the clabber out of a bowl.* Levaillant suspected his mon-

* Thiere der Heimath, vol. i. p 122. Alix. L'esprit de nos bêtes, tells of a hunting dog that deceived his master by pointing at

key, "Kees," of stealing eggs. "So I hid myself one day to watch, when the cackling of the hens proved that they had laid. Kees was sitting on a cart, but as soon as he heard the first cackle he jumped down to get the egg. When he saw me he stood still at once and affected an attitude of great indifference, swayed on his hind legs for a while, and tried to look very artless In short, he used every means to put me off the track and conceal his intentions." * Tame elephants evince remarkable talents in this direction, which are utilized in capturing others. Sir E. Tennent describes a female elephant who excelled in this game. "She was a most accomplished decoy, and evinced the utmost relish for the sport. Having entered the corral noiselessly, carrying a mahout on her shoulders with the headman of the noosers seated behind him, she moved slowly along with a sly composure and an assumed air of easy indifference; sauntering leisurely in the direction of the captives, and halting now and then to pluck a bunch of grass or a few leaves as she passed," etc.† When a pair of wolves fall upon a flock the female often draws the attention of the dogs to herself and lets them chase her while the male seizes the prey ‡ K. Russ says after describing the diseases of parrots: "Some of the cleverest and best-talking birds will sham sickness in a manner that seems incredible Careful scientific observation,

---

imaginary game if he wished to take a direction different from the one followed by the guide

* H. O. Lenz, Gemeinnützige Naturgeschichte, vol i. p 50.

† E. Tennent, Natural History of Ceylon, pp. 181–194. See Romanes, Animal Intelligence. p. 402.

‡ Leroy. Lettres philosophique sur l'intelligence et la perfectibilité des animaux, p. 24.

however, has convinced me of the fact. The bird shows every symptom of disease and lies on the side or stomach, breathing heavily. All this while his master or some one else is in the room, but as soon as he finds himself alone or has reason to think so he appears quite normal and no longer ill. I believe that the explanation of this is that the spoiled pet has noticed that illness excites sympathy, and tender, pitying tones are pleasant to him. Perhaps a slight indisposition or a little pain caused the first complaint, and he has kept it up for the sake of being petted. To cure this unfortunate habit of deception it is only necessary to be a little hard-hearted and not take any notice of the pretended suffering, keeping him as cheerful and busy as possible."*

When we see deception used so effectively to serve practical ends, examples of which are very common, as every student of psychology can testify, it can hardly be doubted that there is in all probability more consciousness of shamming in play than we have any means of demonstrating.

But such a consciousness bears the closest relation to artistic invention, as the following passage from Konrad Lange will show: "If, then, æsthetic performance of children, as well as of primitive peoples, can be proved to have its origin in the play impulse, the next question is whether the same thing is true among animals, and many observations point to an affirmative answer. I will not dilate on this point, only mentioning in passing that many zoölogists believe that certain plays of animals have the character of illusions. Dogs playing with a bone, treat it like prey; cats will do the

* K. Russ, Die sprechenden Papageien, p 396

same with a pebble or ball of yarn. Dogs that are violently excited at the opening of an umbrella or the sight of an empty mouse-trap must experience emotions similar to those of the child at play with his doll or a man at a theatre or admiring a work of plastic art. It is impossible to be certain how far the stimulus to such play is purely sensuous and how much consciousness of illusion is present But it seems to be the general opinion of scholars that there is less of unconscious reflex movement in it than in a recognised illusion play. To establish this would be to gain a very important argument for the significance of conscious illusion in the enjoyment of art; for it is clear that a developmental force that was operative before the evolution of man has a greater claim to be considered the central cause of the gratification that art gives than any number of forces that are not common to the lower animals, however large their part in such gratification may be." *

But before going on we must inquire more particularly what plays this conscious self-deception appears in. Lange, in his fine work Die künstlerische Erziehung, here distinguishes four classes of plays among children—movement plays, sense plays, artistic plays, and rational plays Artistic play is the only one in which conscious self-deception appears, and there it forms an analogue to artistic creation and æsthetic enjoyment. The artistic plays of children are principally dramatic, the child personating its parents or others; even lifeless objects may take part: the table will do for a house, while the footstool is a dog, and so on Other forms, such as the epic play, where stories or pictures

---

* K. Lange, an article in Die Aula, 1895, p. 89.

are acted out, are outside the sphere of the animal psychologist, but he is interested in those directly connected with the imitative arts. Since Lange, both here and in a later article, has found conscious self-deception also in the other arts,* I think it is admissible to include it among the other plays, always with the proviso that consciousness of the sham character of the act is not necessarily present, but may be. The feature common to all animal play is that instinct is manifested without serious occasion. Now, when the animal knows that there is no serious occasion, and yet goes on playing, we have conscious self-deception. It seems to me that this is the case with most of the play of animals, though not with equal certainty; perhaps least of all in imitative play. If we take the dog's play with a stick, for instance, as an example of conscious mock activity, we see that there is no imitation in it, because it is done without a model.†

Glancing over the various kinds of play, can we say that the animal pretends to follow a serious aim when he merely experiments, as when he runs about in a movement play, or springs after a block of wood as if it were prey, or scuffles with his comrade, or amuses himself with building, or treats a young animal of some other kind like a doll, or playfully imitates another, or displays curiosity, or practises his courtship arts? Now, it is evident that the probability of conscious make-believe is a variable quantity in these cases. It seems to be quite certain in the frequently repeated hunting and fighting games, less so in experimentation, move-

* Ibid., p. 21.

† [It is in connection with this question that I have made the suggestion (*Science*, February 26, 1897) stated above in my preface, p. x.—J. MARK BALDWIN.]

ment plays and courtship, and least so in building, curiosity, and imitative play. What makes this difference? Probably the fact that in many plays there is not only sham activity, but also a sham object as well, which, we assume, the intelligent animal recognises as such, while in other cases this is wanting. If we could be certain that apes treat lifeless objects as dolls, this act would be in the foremost rank of illusion plays; if other animals would choose a fixed object as the goal of their races, this too would be most important. But we can not be sure of these things, for speech is wanting to these creatures. The child that puts on his father's hat and says, "Now I am papa," proves that his is not mere instinctive imitation, and that he is conscious of the make-believe, while the monkey that imitates his master has no way of assuring us of the character of his actions. Still less can we ascertain whether the play of masses of animals, which we regard as imitative, is characterized by that absorption of the individual by the mass that is so essential to such play among men.

Be that as it may, there is the strongest probability that the playing animal has this conscious self-deception. The origin of artistic fantasy or playful illusion is thus anchored in the firm ground of organic evolution. Play is needed for the higher development of intelligence; at first merely objective, it becomes, by means of this development, subjective as well, for the fact that the animal, though recognising that his action is only a pretence, repeats it, raises it to the sphere of conscious self-illusion, pleasure in making believe—that is, to the threshold of artistic production. Only to the threshold, however, for to such production belongs the aim of affecting others by the pretence, and pure play has none of this aim. Only love play shows something of it, and in this respect it is nearest to art.

Coming now to inquire into the psychology of the subject yet more closely, we will consider two important points: 1. Divided consciousness in make-believe. 2. The feeling of freedom in make-believe. They are closely connected.

1. *Divided Consciousness in Make-believe.*

A close examination of this conscious self-illusion, which is the highest psychic phenomenon of play, shows that it is a very peculiar condition of mind. I have described it briefly in my work on æsthetics: "I know quite well that the waterfall whose motion I am watching does not feel any of the fury that it seems to show, and yet I remain a captive to the thought that this is so. I see through the illusion, and still give myself up to it." * Something of the same idea, too, is contained in Schiller's words: "It is self-evident that we are here speaking only of æsthetic appearance (Schein) which we distinguish from reality—and yet not logically so, as when one thing is mistaken for another. We like it because it is show, and not because we mistake it for anything else. In other words, we play with it, and this contrasts it with real deception." †

It appears, then, that play, when it rises to conscious self-deception, produces a strange and peculiar

* Einleitung in die Aesthetik. p. 191.

† Ueber die aesthetische Erziehung des Menschen, twenty-sixth letter. See also Kant's weighty utterance: "Nature is beautiful when it appears as art; and art can only be called beautiful when we recognise it as art while it yet appears to us as Nature." (Critique of Judgment, § 45). K. Lange has recently revived this conception of Kant's.

division of our consciousness. The child is wholly absorbed in his play, and yet under all the ebb and flow of thought and feeling, like still water under windswept waves, he has the knowledge that it is only a pretence, after all Behind the sham I, that takes part in the game, stands the unchanging I of real life, which regards the sham I with quiet superiority.*

If now we ask how this phenomenon is related to the other condition of mind known to us, we find that it occupies a position between the ordinary waking state of consciousness and the abnormal conditions of hypnosis and hysteria, which is rather daringly called double personality.†

Many things, it is true, in our waking life suggest a divided consciousness, but the cleft is not so deep as in the abnormal condition. I am not now speaking of the alternation of two psychic existences—that phenomenon is perhaps best illustrated in the everyday life of many heads of families who are unsupportable tyrants at home, while at the club they are the very types of a "jolly old boy"—but I refer rather to simultaneously existing divisions of consciousness, examples of which are not uncommon with us. We may state the case somewhat in this way: It is a formulated scientific fact that a certain economy governs our consciousness. It takes note of but a limited number of the countless physiological stimuli that continually set our brains

* See E. von Hartmann, Aesthetik, vol. ii, p. 59.

† Pierre Janet, L'automatisme psychologique, 1894, p. 132. Max Dessoir, Das Doppel-Ich. 1890. Kant referred to this idea as far back as 1838 in his Träumen eines Geistersehers. Landmann's criticism on The Plurality of Psychic Personalities in One Individual contains much of importance, but his work labours under too sharp a distinction of "cortical" from "subcortical" as used by Meynert.

into activity. We know, further, that human consciousness does not reveal all its store at once, for the mental field of vision is like the optical, in that a part of our store of knowledge is pre-eminent, while all the rest is grouped about the mental view-point (Wundt). I have called this the "monarchical character of consciousness." * But it seems in general, if not always, that the psychic fringe outside of the mental view-point has a certain independence. If we figure the former as a peak, the latter will form neighbouring hills. But how do they arise? In normal cases they are formed from the *débris* of former intellectual operations, which may have been insignificant as psychic phenomena, but are important by reason of their close connection with habits that have become reflex from constant repetition. Thus, when our consciousness becomes full of ideas that are only loosely connected with our habitual I,† it too becomes a neighbouring peak,‡ and so a simple and normal division of consciousness is effected. Condillac recognised this fact and expressed it with the greatest clearness. He says: "When a geometer is intensely occupied with the solution of a problem, external objects continue to act on his senses and the habitual I responds to their impressions. It walks him about Paris, avoiding obstacles while the reflective I is entirely absorbed in the solution." #

* Einleitung in die Aesthetik, p. 3.

† I use the terms "habitual I," "real I," and "apparent I" without the intention of implying actual plurality of personality.

‡ Herein appears also the biological utility of the normal division of consciousness, namely, in that higher intellectual development would be impossible without relative independence of the habitual I.

E. Alix, L'esprit de nos bêtes, p. 587.

In order to make the relation between these two I's, in normal cases, clearer, I cite two commonplace examples from Dessoir: "A friend calls and tells me something that necessitates my going out with him. While he relates the most interesting occurrences I am getting ready to go. I put on a fresh collar, turn my cuffs, fasten the buttons, pull on my coat, get the door key, and even glance in the mirror. All this time my attention is occupied with my friend's narrative, as repeated questions prove. Once in the street, it suddenly occurs to me that I have forgotten the key. I hurry back, look in every nook and corner, and at last feel in my pocket, where, of course, I find it. As I join my friend, he says: 'If you had told me what you wanted, I could have told you that I saw you take the key out of a drawer and put it in your pocket. How can any one be so absent-minded?'" Still more remarkable are the apparently rational automatic movements that we perform mechanically, though they tend to accomplish results that we later acknowledge as our unconscious purpose. An official, for example, sets out in the morning and walks a long distance without once having the idea of his destination enter his mind. But as soon as an acquaintance meets him and inquires why he is out so early, he replies without reflection that he must be at the office." *

Let us now take a simple example from the sphere of hypnotic research. "In the sitting of April 30, 1888," says Dessoir, "the first experiment was made with our principal subject, Herr D——. He received the post-hypnotic suggestion that he should resume the condition as soon as I had clapped my hands seven-

* M Dessoir, Das Doppel-Ich, p. 3

teen times. When he awoke, Dr. Moll engaged him in lively conversation, while I clapped my hands softly, and at irregular intervals, fifteen times. Being asked then whether he had heard my hands striking together, D—— denied it, and, besides, asserted that he did not know what he was to do after the seventeenth clap; but as soon as it sounded he automatically obeyed the order." To this is added: "As D—— had declared that he did not know of the clapping, we put a pencil in his hand with the remark that the hand would write how many times I had clapped. D—— laughed incredulously, went on with his conversation, and did not notice that the pencil wrote '15' with slow strokes —indeed, he would not admit afterward that he had done it." *

I follow up this simple instance with a very remarkable one Pierre Janet made the following experiment with his subject Lucie: During the hypnosis he laid five sheets of white paper on her knee, two of them being marked with a cross. These two he told her she could not see when she awoke. On awaking, she was surprised to see the papers on her lap, and Janet told her to give the sheets to him. She took up those not marked, and declared when asked that there were no more. The marks must have been noted by her "subliminal consciousness" while not suspected by the ordinary one. Janet proceeds: "This supposition was strengthened by complicating the experiment as follows: I put the subject to sleep once more, and placed twenty small slips, all numbered, on her knee. Then I said to her, 'You can not see the papers marked with multiples of three.' When awakened, she showed

* M. Dessoir, Das Doppel-Ich, pp. 18, 22.

the same forgetfulness and the same surprise at finding the papers. I asked her to give them to me one by one; she handed me fourteen, leaving six untouched; these six bore the multiples of three. I am convinced that she did not see them." * In the first of these examples, those taken from everyday life, the division of consciousness is unimportant. When I converse on an interesting theme and at the same time dress myself, brush my hair, wash myself, take a key from the basket, etc., without being able to remember it afterward, it is not at all improbable that my consciousness wandered many times during the talk to the habitual acts. In the hypnotic cases we can not suppose any such glancing off of waking consciousness; there is a deep gorge between the principal and the neighbouring peaks. There are in the same brain two related but independent dynamic complexes.

How is it, then, with conscious self-illusion? Here self-forgetfulness, the losing sight of the habitual I, is, as a rule, more pronounced than in the earlier instance. The child goes about his play very differently from a man engaging in conversation, and many observers testify that playing animals often become blind and deaf to approaching danger, so great is their absorption. But, on the other hand, conscious connection with real life is not so completely broken as in the negative or positive hallucinations of hypnotism, for the sham occupation does not at any time become so absorbing that it can not be changed at will to the reality. Thus it is that division of consciousness as it appears in play forms the medium between the two groups of phenomena which we have considered. Play-

* Pierre Janet, L'automatisme psychologique, p. 277.

ful activity is perhaps more like that resulting from certain dreams. Only a part of the content of many dreams has any relation to the personal consciousness of the dreamer, while the rest appears as something apart, not belonging to him.* For example, Von Steinen dreamed while he was living among the naked tribes of central Brazil that he appeared in European society where all the guests were without clothes. He was rather surprised, but was easily satisfied when somebody told him, " Everybody does it." † Here is a dialogue between the dream I and the waking consciousness which criticises the dream phenomena; the two spheres are so widely separated that they appear as I and you. But dreams that do not allow this are still more like conscious self-deception. We often dream, for example, that we must prepare for some examination that we have already passed, but the waking consciousness quickly interferes with the information that we stood it long ago.‡ If we could show that the dream pictures were not enforced upon our waking consciousness, and that it saw through their shamming and

---

* See H. Siebeck. Das Traumleben der Seele, p 38.

† K. von der Steinen, Unter den Naturvölker Centralbrasiliens, p. 64.

‡ Binet and Féré show that in hypnosis, too, the waking consciousness often rises to the surface. "Every one could probably make some experiments with this dual consciousness by studying his own dreams. Here we see again the relationship between normal sleep and the hypnotic sleep. In general the dreamer is like a somnambulist under a suggested hallucination: he apprehends nothing with certainty: he allows the most palpable absurdities to be perpetrated before his very eyes. But sometimes a remnant of his common sense awakes, and he cries in the midst of the burlesque: 'But this is an impossibility: it must be a dream.'" (Binet and Féré, Le magnetisme animal, p. 107.)

enjoyed the deception, we should be very near the psychological conditions of conscious play.

If, then, in conscious make-believe, in the young dog, for example, that begs his mistress to reach out her foot and then falls upon it with every sign of rage, but never really biting it, the connection between the pretended I and the real I underlying it is preserved in spite of the division of consciousness, the important question to us is concerning the nature of this connection. We might suppose it to be a kind of oscillation from one sphere to the other. Using a commonplace but excellent illustration, it would be like the circus rider who stands with legs wide apart on two galloping horses and throws his balance from one to the other. Lange has expressed the same idea in regard, primarily, to artistic enjoyment, but so as to include play-illusion also; in his book on Künstlerische Erziehung he speaks of the "oscillation between appearance and reality," and regards it as the very essence of æsthetic enjoyment. In a passage on conscious self-deception he goes still further: "Artistic enjoyment thus appears as a variable floating condition, a free and conscious movement between appearance and reality, between the serious and the playful, and since these feelings can never coincide, but must always be at variance, we may adopt the figure of a pendulum. The subject knows quite well, on the one hand, that the ideas and feeling occupying him are only make-believe, yet, on the other hand, he continues to act as if they were serious and real. It is this continued play of emotion, this alternation of appearance and reality, or reason and emotion, if you like, that constitutes the essence of æsthetic enjoyment." *

* Die bewusste Selbsttauschung, p. 22.

I am fully convinced of the truth of the central proposition of this luminous passage, the more since I have come to a similar conclusion in investigating the relation of the sublime to the comic.* But a close examination proves it to be doubtful whether this oscillation between a condition of self-deception and the consciousness of it should be regarded as always a quality of play. Lange seems to me to go too far in making it essential to æsthetic and play enjoyment of all kinds. Self-observation reveals a high degree of satisfaction in long-continued play, during which the real I, as Hartmann justly says, remains quietly in the background and does not assert itself. I do not believe that boys romping together often realize the unreality of their contests while the game is going on; and if we are witnessing the prison scene in Faust our intense enjoyment may last through it all, and our real ego be entirely lost sight of. Only when the curtain falls do we return with a long breath to reality and "come to." Our return to waking consciousness is accomplished more by a sudden leap than by oscillation, and the higher our enjoyment the more rarely do we make the leap.

It will be seen that Lange's proposition is supported less by observation than by logic; he tries to prove his theory of oscillation by the unthinkableness of the reverse. "Since the feeling for reality, on the one hand, and for the apparent can never coincide," he thinks this motion must be regular, but in view of what we know of divided consciousness this seems to me improbable. The examples cited above show that two entirely differ-

* Groos, Einleitung in die Aesthetik, pp. 337 f, 404. So also Kant, Critique of Judgment.

ent psychic processes may run parallel, that there may be a separate subliminal consciousness acting with entire independence. When, for instance, the seventeen hand-claps were registered obediently to post-hypnotic suggestion, the waking consciousness took no note of the count. However, there often seems to be a kind of unconscious connection, like a subterranean wire leading from the subliminal to the waking consciousness, that can not be accounted for by the ordinary change from one state to the other. Even in the deepest absorption, when for a long time there is no recollection of the real ego, we do not substitute appearances for reality. A simple hypnotic experiment of Moll's will illustrate the fact of such a connection: "I told X—— in the hypnotic state that when he awoke he should lay an umbrella on the floor. When he did awake I told him to do what he chose, and at the same time I gave him a folded paper, on which I had written what he would do. He carried out the suggestion, and was amazed when he read the paper He declared that he thought he was doing something this time that had not been suggested." * In a case like this the idea of the act must come over from the subliminal consciousness without the subject's suspecting whence it comes. Emotions, too, may be conveyed in the same way. The subject laughs on awaking, as has been suggested during hypnosis, without knowing that he is obeying a command, and finds some other reason for it.†

* A. Moll, Der Hypnotismus, p. 128.

† That this is not always so is proved by those occasions when the subject bursts out laughing, but afterward knows nothing about it; the feeling is not transferred to the major consciousness. Ibid., p. 120.

Still more remarkable are Binet's observations of hysteria with partial anæsthesia.* For example, the right hand is wholly without sensation, but only so for the waking consciousness, for it grasps a pencil without the patient's seeing or knowing it, finishes a sentence, and even corrects an error intentionally made by the experimenter. There must, then, be a consciousness for which the hand is not anæsthetic.

Many of Binet's experiments indicate that here, too, an unconscious connection exists between the two states of consciousness; hysterical patients may have visual images corresponding to impressions made on the subliminal consciousness. "If, for example, some familiar object, like a knife, is brought into contact with a hand without sensation, the person knows nothing about the form of the knife, about pain inflicted, etc., but all these latent sensations produce their optical counterpart in the sphere of the first consciousness—namely, the visual image of a knife."†

We can attain our object sooner by turning now to E von Hartmann's Aesthetics. I have already referred to his doctrine that the make-believe ego derives æsthetic satisfaction from pretence, while the real ego stands quietly in the background. But besides these apparent feelings we have also real feelings, while we enjoy æsthetic pleasure—namely, our real delight in the apparent.‡ This real pleasure that belongs, as such, to the obscured real ego, now comes over into the sphere of the

* Binet, Alterations of Personality.

† M. Dessoir, Das Doppel-Ich, p. 11.

‡ Von Hartmann, Aesthetik, vol. ii, p. 64. The distinction between make-believe and real feelings is well illustrated by our enjoyment of tragedy, where an unpleasant sham feeling gives real pleasure—namely, æsthetic satisfaction.

play ego. "So it comes about that the happiness produced by æsthetic enjoyment appears as something objective, belonging to the play-scene, and not as a condition of the beholder's soul. It is like a great ocean of bliss on which he floats and moves about at will, having no further influence than to stir it a little, just as a bather gives himself up to passive enjoyment in the encompassing element." *

This appears best in the contemplation of supreme beauty which produces æsthetic pleasure depending on sensuous pleasure. Here sensuous pleasure, an emotion belonging to the real ego and susceptible of physiological explanation, has come, by means of an unconscious connection, into the sphere of make-believe, and lent to the object that divine effulgence which is an attribute of absolute beauty.†

We see, then, that there are many ways and occasions for the use of this unconscious connection between the two states of consciousness, and we must suppose that even in the most absorbing play a constant influence is mutually exerted between them. But what is the character of this influence? It would, of course, be easiest to say that, though the real ego is hidden, it manages to convey its own idea, "This thing is not real," into the sphere of the play-ego. But Lange's objection answers this; he says that while it is possible for a fraud and faith in it to exist side by side in two separate consciousnesses, it is inconceivable that they could be present simultaneously in one and the same sphere. He is right, and if we observe ourselves carefully we will find that it is as far from the truth as

* Von Hartmann, Aesthetik, vol. ii, p. 67.

† *Cf.* Groos, Einleitung in die Aesthetik, pp. 254 f.

is the theory of regular oscillation. Neither in intense artistic enjoyment nor in genuine play does the conscious thought, "This is only a sham," present itself to us. When I said above, in agreement with many others, "I see through the deception, and yet give myself up to it," the actual working of consciousness was, by the bluntness of logical expression, very imperfectly described. For when self-observation assures me that I have given myself up to the illusion, and yet there was no alternation with reality, the logical conclusion arrived at afterward must be that I consciously saw through the sham while I was enjoying it.

The influence proceeding from the real ego is, then, something quite different from this. The fact that in play the apparent does not alternate with the real does not prove that we have a conscious knowledge of the pretence. The solution of the problem seems to me to lie in the simple fact that consciousness of the apparent is from the outset, and, in spite of all similarity, quite different from consciousness of the real; and I find the final ground for this difference in nothing less than the fact that we recognise ourselves as the cause of the pretence.* This brings us again to the idea of joy in being a cause; the real I feels itself to be the originator of the make-believe images and emotions which it calls forth voluntarily, and this feeling of being a cause glides over unconsciously to the world of illusion and gives to it a quality not possessed by reality. Reality oppresses

* In my Einleitung in die Aesthetik (p. 82) I have emphasized this, and have encountered the criticism of having expressed opinions concerning the essence of the soul that are not susceptible of proof. Here we are only concerned with the fact that we undoubtedly feel ourselves to be the cause; whether we really are so is indifferent for the purposes of a psychology of play.

us with a sense of helplessness, while in the world of illusion we feel free and independent. There is no need to say, "This is not real," for every idea and feeling that forms part of the illusion bears the stamp *ipse feci*, and can not be confused with reality. Only when the consciousness of being a cause leaves the obscured real ego does such confusion take place, and then the mind's condition ceases to be playful and becomes pathological

Let us take an instance of the dangerous trifling with the emotional nature, so common in our day, when a nervous and excitable person arouses his emotions without any real cause. Marie Baschkirtzew writes at the age of thirteen years: "Can it be true? I find everything good and beautiful, even tears and pain. I love to weep, I love to despair, I love to be sad. I love life in spite of all, I wish to live. I must be happy, and am happy to be miserable. My body weeps and cries, but something in me that is above me enjoys it all." Can we suppose that the unhappy young girl had the clear idea amid her storm of emotion, "These feelings have no real cause," and that she created from this knowledge this strange ecstasy of pain? Is it not much more probable that this feeling was wanting during the rush of emotion, and that what produced the ecstasy was the feeling of pleasure in being a cause that came over from the real I, the feeling that all this agitation was not contrary to her will but produced by herself; in other words, the feeling of being active and not passive, the feeling of having produced a sublimated kind of reality through her own psychic activity? Only afterward comes the logical formulation, "My sorrows, my joys, and my cares have no existence"—an idea that is not present in the first gush of feeling, and if it were, would only increase the pain instead of changing it into a subject for rejoicing.

I believe, therefore, that in genuine absorbing play the oscillation from appearance to reality is an unnecessary as well as an improbable hypothesis. The idea unrecognised by consciousness gliding over from the real ego, that the whole world of appearance depends on ourselves, that we create it from material within us,* is sufficient to prevent our mistaking the make-believe for reality, without, however, making it necessary for us clearly to hold the difference in mind.† This conclusion brings us to a second point, which we may now consider, finding in it a more definite answer to the question.

2. *The Feeling of Freedom in Make-believe.*

Connecting the idea of freedom with that of make-believe brings us back to Schiller. There are two kinds of temperament belonging to genius. The one strives for what is attainable, the other for what is not. Schiller says: "The one is noble by reason of attainment, the other in proportion as he approaches infinite greatness." He himself belongs to the second class; he with Michelangelo and Beethoven are types of the eternally striving and struggling genius straining for the unattainable, in whom the artist's gift is nourished by

---

* This is why we are proud of the capacity for such creation. It is a kind of joy in being able.

† In support of this position I appeal again to the tragic drama. When feelings produced by inner imitation, and voluntarily called forth, become so painful as to counterbalance the pleasure derived from æsthetic satisfaction, we call in the help of our knowledge of its unreality to dampen the ardour of our emotions. But so long as we are in full æsthetic enjoyment we do not think of this until the play is over, though it is responsible for half our pleasure.

"gleams from the lamp of life itself." Schiller's youthful philosophy disclosed this principle of his nature. Above the actual world, with its suffering, above "this dream of warring frogs and mice," this life of frivolity, a lofty spiritual world rises in glorious perfection, to which he ascribes the fulfilment of every ideal of love, friendship, joy, and freedom. But this beautiful world, already threatened by Voltaire, vanished before the chill breath of the destroyer of ideals—Kant. Schiller expressed his pain in the loss of the ideal in his Gods of Greece. That noble blooming time of Nature represents to him the flowering of his own youthful idealism; and when he bewails "all the fair blossoms falling before the blasts of winter," much that is personal is hidden in the words.

The ideal is only a dream, a beautiful chimera, but need not, therefore, be lost to us, for we may still enjoy the ideal in play; and with this conception, the poet rises to new flights which open the classic period of his creation.

It is necessary to apprehend this fact clearly in order to understand the great ethical power of Schiller's Æsthetics, which is for him not merely a new intellectual discipline, but, above all, a new victory of ethical personality. Being denied metaphysical ideals, he directs his whole ethical force to the realm of beauty, and feels that in virtue of his art he is a priest of humanity, whose honour is intrusted to his care. In beautiful unreality he finds again all that he dreamed in youth, harmony of feeling and impulse, happiness, freedom, and the highest perfection of mankind. His metaphysical idealism comes back to him in the form of æsthetic idealism.

Inquiring more closely into the nature of this æsthetic idealism, we find that it culminates in the feeling of freedom; when indulging in it a man is free—that is to say, he is wholly human only when he plays, for there is no real freedom in the sphere of experience. In real life the man is a plaything of opposing forces. On the animal side of his nature, the sensuous, he is restrained by Nature's laws, while reason forces him to obey imperious moral mandates, and a perfect reconciliation of these forces is impossible. "Between pleasure of the senses and peace of mind man has but a sorry choice." Only in playing and indulging in beautiful dreams can a man find relief from this contention. Schiller expressed this conviction when he was in Mannheim, as far back as 1784. "Our nature," he says, "alike incapable of remaining in the condition of animals and of keeping up the higher life of reason, requires a middle state, where the opposite ends may unite, the harsh tension be reduced to mild harmony, and the transition from one condition to the other be facilitated. The æsthetic sense, or feeling for beauty, is the only thing that can fill this want." And what is the governing idea in this middle state? "This: to be a complete man." * By reducing in his play the harsh tension to mild harmony he relieves himself of the double law of Nature and Reason, raises himself to a state of freedom, and so first attains his full humanity. The result achieved in play is "the symbol of his true vocation." †

Schiller says: "The sensuous impulse must be expressed, must attain its object; the form impulse expresses itself and produces its object; but the play im-

* Schiller, Die Schambuhne als eine Moralische Anstalt betrachtet.

† Aesth. Erziehung, fourteenth letter.

pulse strives to receive as if itself had produced the object, and to give forth what sense is labouring to absorb. The sensuous impulse excludes from its subject all self-activity and freedom; the form impulse excludes all dependence and passivity. But the exclusion of freedom is physical necessity, and the exclusion of passivity is moral necessity. Both impulses constrain the soul, one by natural laws, the other by moral laws. The play impulse, then, uniting them, affects the mind both morally and physically, lifts it above both accident and necessity, and sets man free, physically and morally." * "The term 'play impulse' is justified by the usages of language, which signifies by the word play (*Spiel*) all that is neither contingent subjectively or objectively, nor yet either internally or externally compelled. Thus the mind, by beholding the beautiful, is placed in a happy mean between law and necessity, and relieved from the oppression of either, because it is divided between the two." †

Passing over Schiller's hair-splitting method of establishing the equilibrium between the two opposing impulses—which he suspended like two equal weights in a balance, being still controlled by the old theory of faculties—and without elaborating these ancient ideas, we will rather attempt to translate them into modern psychological language. First, then, Schiller is perfectly right in designating the feeling of freedom as the highest and most important factor in the satisfaction derived from play, and further in finding it closely related to the feeling of necessity. We feel free although we are compelled; this is indeed the very essence of

* Aesth. Erziehung, fourteenth letter.

† Ibid., fifteenth letter.

play. We are compelled, for sham occupation is related to the hypnotic condition in that it treats mere appearance as if it were reality. The make-believe I follows all the turns of playful activity, yielding obedient service to the intellectual and emotional stimuli which they evolve, and yet this compulsion is not like that which oppresses us in actual experience, for the fact is always present to our consciousness that we are the creators of this world of appearances. "The reality of things," says Schiller, "is inherent in them, the appearance of things is man's affair, and the state of mind that is nourished by appearance takes more pleasure in its own activity than in anything that it receives." * We are compelled, because we are under the power of an illusion, and we are free because we produce the illusion voluntarily. Indeed, it may safely be said that we never feel so free as when we are playing.

Apart from all transcendental considerations, free activity, regarded from a psychological standpoint, depends on our ability to do just what we wish to do, and on no other ground; this is the positive side, and the negative side is that we have the conviction that we can abstain from the act at any moment that pleases us. The popular idea is correct in calling a man free when he does and leaves undone what he chooses, for the feeling of being at liberty consists in regarding ourselves as the arbiters of our own destiny. Whatever error the theoretical metaphysician may think it necessary to combat in this statement, it remains a psychological fact that we do have such a feeling, and that it is of incalculable practical significance. Let us see in what it consists. We feel ourselves to be absolute causes—

* Aesth. Erziehung, twenty-sixth letter.

that is to say, we feel ourselves to be governed entirely by ourselves, by our present will. No "not I" seems to us to influence either our present object or the idea of our former or future experience; we seem to be divided from the all-powerful causal nexus pervading the ages, and to be at liberty to fulfil our present desires unencumbered by circumstances or consequences. We seem, as Kant expresses it, to begin a causal series "self-originated and elemental." *

The feeling of freedom is undoubtedly heightened by our conviction that we can desist from an act at any moment. "I am still free" is the same as "I can yet turn back." Here, also, freedom is identical with being an absolute cause, for if I were able only to set an act on foot, but not go on with it, my freedom would vanish as soon as my causality ceased. So the struggle for liberty turns out to be the highest psychic accompaniment of the struggle for life. The instinctive propensity of all living creatures to preserve their independence, to shake off every attempt on individual liberty, culminates in the effort after intellectual liberty. The joy of freedom is the sublimest flight of that pleasure in being a cause, which has occupied so much of our attention.

But where can the feeling of freedom be purer or more intense than in conscious self-illusion in the realm of play? In real life we are always in servitude to objects and under the double weight of past and future. These objects, intelligent and otherwise, for the most part oppose our wills or assume authority over us. Care for the future torments us and robs us

---

* Critique of Pure Reason (p. 435 of Kehrbach's German edition).

of our freedom of action. The past, which no more belongs to our living ego, is riveted to us with iron bolts so that we can not escape from it. And where in real life is the feeling that we always might turn back—might step out of the causal series? Perhaps our resolution seems to be free, but as soon as stern realities beset us we fall again under the resistless causal nexus of the universe, and no power on earth can send back the arrow that is loosed from the bowstring. We may well suppose that it was under bitter experience of the inevitableness of necessity that Schiller described Wallenstein's condition with such force of genius. Perhaps the power of the "not I" over the "I" has never been more tragically set forth than in that great monologue, where we see the unlucky stars depriving the hero of his freedom:

"Is it possible?
Is't so? I *can* no longer what I *would*?
No longer draw back at my liking, I
Must *do* the deed because I *thought* of it?
. . . .
I but amused myself with thinking of it.
The free will tempted me, the power to do
Or not to do it.—Was it criminal
To make the fancy minister to hope?
. . . .
Was not the will kept free? Beheld I not
The road of duty close beside me—but
One little step and once more I was in it!
Where am I? Whither have I been transported?
No road, no track behind me, but a wall
Impenetrable, insupportable,
Rises obedient to the spells I muttered
And meant not—my own doings tower behind me.
. . . .
Stern is the on-look of Necessity.
Not without shudder may a human hand
Grasp the mysterious urn of destiny.

My deed was mine, remaining in my bosom:
Once suffered to escape from its safe corner
Within the heart, its nursery and birthplace,
Sent forth into the Foreign, it belongs
Forever to those sly, malicious powers
Whom never art of man conciliated." *

"Stern is the on-look of necessity," says Wallenstein, and "Life itself is stern," cries Schiller in the prologue to the same drama. But—"art is brighter and more cheerful." The effect of play is brightness and freedom—so much so that we may say, in real life there is freedom only so long as serious activity is not yet begun—that is, while the man still plays with conflicting motives. What do the advocates of indeterminism mean by freedom? It is to them the ability to choose among various motives; but this choice is nothing but a play in which the man represents to himself now this, now that motive as realized; it is a conscious self-illusion. And only when he has indulged in it does he feel, after the decision is made, that he has acted freely. Wallenstein's monologue has a special interest in this connection; for, since he found pleasure in amusing himself with the mere thought of royalty and delighted in the illusion, it is clear that for him the feeling of freedom consisted in this play of motives. The word for play in most languages signifies only a pleasurable condition, but the old German word *Spilan* means a light floating movement †—that is to say, free activity—giving to the modern word *Spielen* a primary significance which bears out our analysis. Freed from the causal nexus of the world's events, play

* From Coleridge's English version of The Death of Wallenstein.

† M Lazarus, Ueber die Reize des Spiels, Berlin, 1883, p. 19.

is a world to itself, into which we enter voluntarily and come out when we will. There we seem freed from necessity because in conscious self-illusion we feel ourself to be an absolute cause.

We are now approaching the end of our inquiry. The joy in being a cause having culminated in the highest and most refined of pleasurable feelings—namely, in that of liberty—we find here the deep significance of that division of consciousness which occupied us in the last section. The difficulty of explaining it consists in the fact that in play we take appearance for reality, and still do not confuse it with the actual. In many cases the leaping over of our consciousness to the real I is conceivable, but in the most intense enjoyment this off-shooting of consciousness does not take place, and we must suppose an unconscious connection between the real and play egos that obviates the necessity for this alternation. We have found such a connection in the feeling of being a cause without going into the nature of these psychic adjuncts of make-believe. This is now the place for such an inquiry.

I have throughout this whole treatise spoken not of the *idea* but of the *feeling* of being a cause. A conscious idea that we ourselves produce the appearnce is as little supposable during intense enjoyment as the idea, "This is only a pretence." What glides over from the real I, and is recognisable by self-observation, is only the feeling of pleasure arising from the consciousness of being a cause and culminating in the feeling of freedom. There are, empirically speaking, no pure feelings that can be distinguished from ideas as such, no abstract pleasure or pain.* Feeling is always, in its finer manifestations, the

* *Cf.* Lehmann, Hauptgesetze des Gefühlslebens, p. 16.

product of intellectuality, but the intellectual elements are latent and are manifested only in the shading that they impart to the emotions. So is it in the case we are considering. The consciousness of the obscured real ego that has produced * the whole illusion, and so created a free world of appearance above the causal nexus of reality, does not appear conspicuously in the feeling of freedom that oversteps the bounds of the apparent world, but does impart to it a character that distinguishes it from all other pleasurable feelings. This characteristic seems to me to form the barrier that prevents our confusing the make-believe with the real.

The artist always employs some means to prevent such confusion—the frame, for example, in painting and the pedestal for a statue. Theodor Alt † includes all such means under the general name of "negative effects," while Conrad Lange calls them "illusion-destroying effects." ‡ In play the feeling of freedom subjectively performs the office of these objective means. It gives the whole world of appearance a special colouring, distinguishing it from everything that is real, and rendering it impossible that even in our utmost absorption we should ever confuse the make-believe with the real. As in æsthetic enjoyment, the real pleasure in beholding—which is, after all, only a special case of our general principle—steps over into the apparent

* Th. Ziegler says of the feeling of freedom: "But what is the nature of this feeling? Only 'that all my actions proceed from myself, that I am the cause of them'; it is closely related to the feeling of power, one side of it, so to speak, isolated, strengthened, and generalized, and belonging to the whole ego; just as in the feeling of dependence, on the contrary, the essential thing is subjection of the ego as a whole." (Das Gefühl, p. 293.)

† Alt, System der Kunst, p. 23.

‡ Die bewusste Selbsttäuschung, p. 20.

world and changes it into a better and higher one, so in conscious play the whole sham occupation is transformed by the feeling of freedom into something higher, freer, finer, and more luminous, which we can not confuse with the realities of life. The feeling of freedom, then, is the subjective analogue to the objective "destroyers of illusion." Life is earnest, art is playful.

I wish to append to this concluding chapter a brief note. Should a question be raised as to the nature of the artistic production whose germ is present in the animals, the following may serve as an answer: First, there is the commonest of all kinds of play, experimentation, which, with its accompanying joy in the possession of power, may be regarded as the principal source of all kinds of art. We have also found, in the excitement created by musical sounds, an approach to human art. We recall the monkey that took great pleasure in striking on hollow objects. From experimentation in general three specialized forms of play arise, analogous to the human arts, and their differentiation leads us to the three most important principles of the latter. They are courtship, imitation, and the constructive arts, and the three principles involved are those of self-exhibition, imitation, and decoration. These principles are expressed in art as the personal, the true, and the beautiful. There is no form of art in which they are not present together, though one usually dominates, while the others are subsidiary. This is evident even in the animal world. The bird that adorns his nest imitates the example of others, and expresses his personality in the work. The bird that mimics another often effects an improvement in his own song, and indulges in self-exhibition; and the bird that displays his skill to admiring females does not fail to employ the principles of imitation and decoration. So we find in animals, and especially in birds who, though so distantly related to us, seem by reason of their upright carriage more near, a certain analogy to our own system of arts; indeed, in the simplest phenomena displayed in the animal world we recognise an important suggestion as to the solution of the vexed question of the proper natural division of human arts. The recognition of the three fundamental principles, which are, however, held together to the single one of experimentation, seems to me a gain, as opposed to the one-sidedness of many investigators. This relationship points directly to the fact that all forces efficacious in artistic production are referable to the central idea of play, and therefore to an instinctive foundation. The following table will make this clear:

PLAY.

Experimentation.
(Joy in being able.)

(Pretence : conscious self-deception.)

Self-exhibition.	Imitation.	Decoration.
The personal.	The true.	The beautiful.
With animals. { Courtship arts.	Imitative arts.	Building arts.
With man. { Dance with excitement. Music. Lyric poetry.	Imitative dance. Pantomime. Sculpture. Painting. Epic poetry. Drama.	Ornamentation. Architecture.

EDITOR'S APPENDIX
ON ORGANIC SELECTION.*

IN certain recent publications † an hypothesis has been presented which seems in some degree to mediate between the two current theories of heredity. The point of view taken in these publications is briefly this: Assuming the operation of natural selection as currently held, and assuming also that individual organisms through adaptation acquire modifications or new characters, then the latter will exercise a directive influence on the former quite independently of any direct inheritance of acquired characters. For organisms which survive through

* See pp. 64, 65, above. This appendix reproduces a communication made to Science (April 23, 1897) and Nature (April 15, 1897), slightly revised.

† H. F. Osborn, Proceedings of the New York Academy of Science, meeting of March 9 and April 13, 1896, reported in Science, April 3 and November 27, 1896; also American Naturalist, November, 1897. C. Lloyd Morgan, Habit and Instinct, October, 1896, pp. 307 ff., also printed in Science, November 20, 1896. J. Mark Baldwin, discussion before the New York Academy of Science, meeting of January 31st, reported in full in Science, March 20, 1896, also American Naturalist, June and July, 1896; also see other references given above, p. 64. The following brief statement was prepared in consultation with Principal Morgan and Professor Osborn.

adaptive modification will hand on to the next generation any "coincident variations" (i. e., congenital variations in the same direction as adaptive modifications) which they may chance to have, and also allow further variations in the same direction. In any given series of generations, the individuals of which survive through their susceptibility to modification, there will be a gradual and cumulative development of coincident variations under the action of natural selection. The adaptive modification acts, in short, as a screen to perpetuate and develop congenital variations and correlated groups of these. Time is thus given to the species to develop by coincident variation characters indistinguishable from those which were due to acquired modification, and the evolution of the race will proceed in the lines marked out by private and individual adaptations. It will appear as if the modifications were directly inherited, whereas in reality they have acted as the fostering nurses of congenital variations

It follows also that the likelihood of the occurrence of coincident variations will be greatly increased with each generation, under this "screening" influence of modification; for the mean of the congenital variations will be shifted in the direction of the adaptive modification, seeing that under the operation of natural selection upon each preceding generation variations which are not coincident tend to be eliminated.*

Furthermore, it has recently been shown that, independently of physicial heredity, there is among the animals a process by which there is secured a continuity of social environment, so that those organisms which are

* This aspect of the subject has been especially emphasized in my own exposition. American Naturalist, June, 1896, pp. 147 ff.

born into a social community, such as the animal family, accommodate themselves to the ways and habits of that community. Prof Lloyd Morgan,* following Weismann and Hudson, has employed the term "tradition" for the handing on of that which has been acquired by preceding generations; and I have used the phrase "social heredity" for the accommodation of the individuals of each generation to the social environment, whereby the continuity of tradition is secured.†

It appears desirable that some definite scheme of terminology should be suggested to facilitate the discussion of these problems of organic and mental evolution; and I therefore venture to submit the following:

1. *Variation:* to be restricted to "blastogenic" or congenital variation.

2. *Accommodation:* functional adaptation of the individual organism to its environment. This term is widely used in this sense by psychologists, and in an analogous sense by physiologists ‡

3. *Modification* (Lloyd Morgan): change of structure or function due to accommodation. To supercede "ontogenic variations" (Osborn)—i. e., changes arising from all causes during ontogeny.

4. *Coincident Variations* (Lloyd Morgan): variations which coincide with or are similar in direction to modifications.

---

* Introduction to Comparative Psychology, pp. 170, 210; Habit and Instinct, pp 183, 342.

† Mental Development in the Child and the Race, first edition, January, 1895, p. 364; Science, August 23, 1895; more fully treated in Social and Ethical Interpretations, 1897, chap. ii.

‡ It may be thought that "individual adaptation" suffices for this; but that phrase does not mark well the distinction between "accommodation" and "modification." Adaptation is used currently in a loose general sense.

5. *Organic Selection* (Baldwin): the perpetuation and development of (congenital) coincident variations in consequence of accommodation.

6. *Orthoplasy* (Baldwin): the directive or determining influence of organic selection in evolution.*

7. *Orthoplastic Influences* (Baldwin): all agencies of accommodation (e. g., organic plasticity, imitation, intelligence, etc.), considered as directing the course of evolution through organic selection.

8. *Tradition* (Lloyd Morgan): the handing on from generation to generation (independently of physical heredity) of acquired habits.

9. *Social Heredity* (Baldwin): the process by which the individuals of each generation acquire the matter of tradition and grow into the habits and usages of their kind.†

J. MARK BALDWIN.

---

* Eimer's "orthogenesis" might be adopted were it possible to free it from association with his hypotheses of "orthogenic" or 'determinate" variation, and use-inheritance. The view which I wish to characterize is in some degree a substitute for these hypotheses.

† For further justification of the terms "Social Heredity" and "Organic Selection," I may refer to the American Naturalist, July, 1896, pp. 552 ff.

INDEX.

THE END.

Lightning Source UK Ltd.
Milton Keynes UK
UKHW022317060223
416579UK00001B/415